Dibakar Pal

POETIC LICENCE (A collection of creative nonfictions)

Dibakar Pal

POETIC LICENCE (A collection of creative nonfictions)

JustFiction Edition

Imprint

Any brand names and product names mentioned in this book are subject to trademark, brand or patent protection and are trademarks or registered trademarks of their respective holders. The use of brand names, product names, common names, trade names, product descriptions etc. even without a particular marking in this work is in no way to be construed to mean that such names may be regarded as unrestricted in respect of trademark and brand protection legislation and could thus be used by anyone.

Cover image: www.ingimage.com

Publisher:
JustFiction! Edition
is a trademark of
Dodo Books Indian Ocean Ltd. and OmniScriptum S.R.L publishing group

120 High Road, East Finchley, London, N2 9ED, United Kingdom
Str. Armeneasca 28/1, office 1, Chisinau MD-2012, Republic of Moldova, Europe
Printed at: see last page
ISBN: 978-620-0-10570-7

PREAMBLE

Anybody can write, since everybody is writer.

Dibakar Pal

POETIC LICENCE

(A collection of creative nonfictions)

DEDICATION

IN MEMORY

OF

LOST FACES

TABLE OF CONTENTS

OF METICULOUS

ABSTRACT

If casual attitude serves the purpose then, who is so foolish to be a serious one? Seriousness i.e. meticulousness deprives one from natural enjoyment. Only a judicious person prefers and practises to be a meticulous master. Common people seldom care for detail. They are happy with retail. So a lay man should not be entrusted with detail work. They are fit for casual job, seldom a meticulous one.

KEYWORDS: Meticulous, careful, attention, extreme, excess, detail, accurate, exact

INTRODUCTION

Creative writing is based more on manifestation rather than on expression. It does not inform, rather it reveals. So it bears no reference. The best creative writing is critical, and the best critical writing is creative. This article is an outcome of thinking about creative writing meant for a general readership. As such, I have adopted a free style methodology so that everyone can enjoy the pleasure of reading. As you might know, Francis Bacon (1561-1626), the immortal essayist, wrote many essays namely 'Of Love', 'Of Friendship', 'Of Ambition', 'Of Studies', and so on. The multiple-minded genius correctly pointed out that all the words of the dictionary can be used as themes for essays. But little has been done since his death to continue or finish his monumental task. Bacon's unique individual style of presentation ignited my imagination and encouraged me to write creative essays as a method of relieving a wide range of emotions through catharsis.

ARTICLE

Meticulous means extremely or excessively careful about minute details. It is scrupulously careful. It is popularly over careful e.g. She is meticulous about her appearance. It implies finicky e.g. meticulously clean. It is giving or showing great care and attention to detail e.g. a meticulous worker/researcher. It is very careful and precise e.g. a meticulously planned schedule. It implies accurate e.g. do meticulous and painstaking research. It means exact. It means JIT i.e. just in time.

To take care is good. Caring parents are good. They are valued much by the society. The children prosper appropriately. They reciprocate their respect to their parents while they are established. Prodigal children are the outcome of negligence by the parents. As such they blame their parents later on while they face misfortune.

Too much of everything is bad. Similarly, too much care is bad than no care at all. It is an obsession that causes annoyance to the concerned child and observer as well. This extreme care is very harmful for the development of the personality of the child. It is a hindrance for full blooming of the tender soul. The child becomes fully dependent. It cannot move alone. It can only rise, like a tree, with the tender touch of the affectionate gardener of the garden. Later on, when there is no guide then the person stands still like a statue. It cannot tackle the hard reality. It becomes handicapped. Affection renders the parents blind and blunt as well. So they become the prey of meticulous care.

One tea spoon full of sugar is sufficient for a cup of tea. Now, if excess amount of sugar is poured then it is no more tea at all. And one cannot take that tea. Over dose causes simply wastage. Over dose invites danger always in medication. Here the patient becomes ill more than early recovery. It increases pain rather than relief. Here, meticulous care should be taken for proper medication and optimum amount of medicine, as well.

Familiarity gives birth to friendship. But too much familiarity brings contempt. One may be meticulous. One may insist one to be meticulous for the benefit of the concerned person. But one should not pressurise. For, all cannot be meticulous. Then either, anger or frustration will come back in return.

Meticulous research is good. Meticulous inquisitiveness is also good. Meticulous quarrel is bad. It pollutes the soul. It wastes both time and energy. The parties involved lose both who were close to each other in the past. The younger generations get negative feedback from these dirty events. It affects their minds. They cannot free themselves from these cloudy affairs.

Instruction may not be meticulous always. But lucid instructions pave the way for smooth work by the novice or less expert workers. An overall guidance is required for the implementation purpose. Freedom should be given to the implementer. It is better to leave the job on one's judgement. Then the outcome becomes well. If the implementer is wise then he should be given independence for execution. In fact, a meticulous worker does not need detail guidance; rather he can act as a guide. If the worker is dull, he should be guided, since meticulous outcome cannot be expected from such an inexperienced guy.

Meticulousness is a habit. Some communities have this quality. Again, some communities lack this trait. Thus all cannot be meticulous person. It needs patience. Patience is alias and akin to physical pain. Very few people can bear that pain. This answers why we see very few meticulous persons around us.

Common people are very superficial. They cannot stick to any matter for long. Due to superficial knowledge a person fluctuates. In depth knowledge acts as brake. A meticulous person avoids casual persons. He is serious and sincere. Such a sincere person is rigid and overzealous, meticulous, over conscientious, inelastic. He reads the proofs with meticulous care. He knows that any lackadaisical attitude will cause the total publication project quite frustrated.

A meticulous student shines in life. An inattentive student cannot write all the answers in any examination. A meticulous student writes full answers. Due to meticulousness his knowledge becomes thorough. A meticulous student does not write different things. He writes differently. His knowledge is perfect. Through this perfection he attains success from cradle to grave.

An inattentive student fails in the examination. As such, misfortune dogs him wherever he goes. Thus he fails everywhere till he breadths his last. Man realises his callousness in his autumn of life. So childhood needs guidance. All cannot be autonomous. Without guidance a person becomes diverted. Such a diverted genius becomes a misguided missile. Then he becomes either dangerously brilliant or brilliantly dangerous or both simultaneously.

All the works of the society is not critical. So, expert workers are not required always. Most of the stereo-typed jobs are done by casual workers who are low waged labourer. In case of critical job meticulous expert is required. An expert is a high paid employee. Such a master mind prepares the master plan. Then the master plan is sub-divided into small modules. The junior workers are engaged to work on these modules. All the modules may or may not be identical. The lower level employees cannot guess the master plan. The master plan is completed after assembling all these modules. This type of module based work is common in computer operation. Since, the master plan remains unknown to the lower level or medium level managers the secrecy of the organisation remains protected.

Meticulousness is an art. Art is not for all. Also all is not artist. This answers why we see very few meticulous persons around us. All cannot be meticulous master. It requires patience. It demands talent. It needs reserve personality. Also all need not to be meticulous worker. All should not be meticulous employee. Only the boss should be meticulous person. The employees will be casual worker who will follow him blindly. In case of stereo-type job blunt workers are more suitable than the blind workers. Thus the head of boss and hands of the labourer make the complete man.

Boss should not be extrovert. Then secrecy of the project thereby secret policy of the organisation will be disclosed to the opponent. As such, during selection of meticulous worker extrovert or frank candidate is not selected. Meticulous worker becomes boss in course of time. An extrovert person is popular. Everybody loves him. Nobody is afraid of him. He cannot be rigid. Rigidity is not his personality. Through tenderness administration cannot be run. A Good Samaritan is suitable for any charitable institution, but seldom for business organisation. Rough and tough are chief ingredients for financial gain of any profitable institution.

CONCLUSION

If casual attitude serves the purpose then, who is so foolish to be a serious one? Seriousness i.e. meticulousness deprives one from natural enjoyment. Only a judicious person prefers and practises to be a meticulous master. Common people seldom care for detail. They are happy with retail. So a lay man should not be entrusted with detail work. They are fit for casual job, seldom a meticulous one.

REFERENCES
No reference, since the present article is an outcome of Creative Nonfiction Writing.

OF LISTENING

ABSTRACT

In case of love both the lovers will speak and listen to each other. Both the eloquent lovers are made for each other. But an eloquent and silent partner is mad for each other for being opposite in characteristic nature and behaviour. Both these categories are good lovers. Now, if the partners are both silent then they are great lovers. They say where goodness ends, greatness begins. Silence is more eloquent than speaking anything. Silence beckons. It implies consent. Both enjoy unfathomable voice of listening through eyes.

KEYWORDS: Listen, hear, attention, obey, follow

INTRODUCTION

Creative writing is based more on manifestation rather than on expression. It does not inform, rather it reveals. So it bears no reference. The best creative writing is critical, and the best critical writing is creative. This article is an outcome of thinking about creative writing meant for a general readership. As such, I have adopted a free style methodology so that everyone can enjoy the pleasure of reading. As you might know, Francis Bacon (1561-1626), the immortal essayist, wrote many essays namely 'Of Love', 'Of Friendship', 'Of Ambition', 'Of Studies', and so on. The multiple-minded genius correctly pointed out that all the words of the dictionary can be used as themes for essays. But little has been done since his death to continue or finish his monumental task. Bacon's unique individual style of presentation ignited my imagination and encouraged me to write creative essays as a method of relieving a wide range of emotions through catharsis.

ARTICLE

Listen is to make an effort to hear somebody/something. For example: We listened carefully but heard nothing. Listen! What's that noise? You are not listening to me/to what I am saying! Listen carefully and you will hear something moving about in the garden.

Listen is to pay close attention for the purpose of hearing. It is to give heed or pay attention to what is said. It is to obey. For example: I have told him repeatedly, but he does not listen. It is to wait attentively for a specific sound e.g. to listen for the telephone.

Listen is to make a conscious effort to hear. It is to attend closely, so as to hear. It is to follow advice. It is to give ear or hearken. It is to attentively hear a person speaking.

It is to give attention with the ear to person or sound or story e.g. listened to my story. It is to take notice of. It is to respond advice or a request or to the person expressing it. It is to yield to, person furnishing, temptation or request.

It is to allow oneself to be persuaded by somebody making a suggestion, giving advice, etc. For example: I never listen to i.e. believe what salesmen tell me. I should never have listened to him. Why won't you listen to reason? I warned her not to go but she would not listen. I tried to persuade him, but he would not listen.

Listen out for something is to be prepared to hear something. It is to seek to hear sound etc. by waiting alertly for it. For example: Please listen out for the phone while I am in the garden. We listened out for the least noise.

Listen in on/to something is to listen to a wireless/radio broadcast e.g. listen in to the BBC World Service. It is to overhear intentionally a message intended for another. It is to eavesdrop. It is to tap a private conversation, especially, one by telephone. In archaeology, it is to give ear to. It is to hear. For example: She loves listening in on other people's conversations. The criminals did not know the police were listening in e.g. by recording their telephone calls.

Listen is an act of listening. For example: Have a listen and see if you like it.

Listener is a person who listens or hearkens e.g. a good listener i.e. somebody who can be relied on to listen with attention or sympathy. It is a person receiving broadcast radio programme.

Good listener is one who habitually listens with interest or sympathy.

Listenable is that can be listened to, especially with pleasure.

Listening post, in military, is an advanced, concealed position near the enemy's lines, for detecting the enemy's movements by listening. It is a post where men are stationed to hear what the enemy is doing. It is any strategic position or centre for securing information or intelligence. It is any source of secret information. It is a station for intercepting electronic communications. It is a place for the gathering of information from reports etc.

When everybody speaks at a time then it is quarrel. If the parties involved are all equally powerful then it is endless voyage. Everybody tries to overtake the other through shouting or fighting or with both simultaneously. In such a noisy situation if one keeps silent and hears the opponent patiently then even the stronger rival opponent stops shouting which is not otherwise possible. The stronger rival stops thinking that he has established his superiority. Also he is happy after serving and satisfying his ego.

Someone is afraid thereby eager more to speak rather than listening lest the opponent wins. The mightier speaks and compels the weaker to hear. A wise man does not compel the weaker one to surrender. The learned knows that if the weak person of today if becomes stronger in future then he will take revenge. As such they say look before you leap.

In any court there is no place of shouting. Everybody enjoys opportunity of being heard. It is the place of listening. A good listener can argue logically with better convincing ingredients. He always wins. It is the essence of democracy. Here lies the triumph and importance of listening.

Man prefers to hear favourable words. Sweet song or melody, though tender in nature, compels a person to listen without applying any force. It is spontaneous and automatic. No enforcement of law is required in this regard.

In any gathering if the speaker is good then the audience hears with patience. If the speaker cannot speak well logically then the audience leaves the place. Further if the topic is beyond knowledge or below than average then the audience also leaves the speaker.

If a noble laureate offers speech and if there is huge gathering then the society is educated. Further, if there is a program of Hollywood heroine and if there is much gathering then the society is uneducated.

Now if the aforesaid two programs occur simultaneously then from the gathering of people one can judge and gauge the culture of the concerned citizens.

As such the speaker must speak as per status of the mob. It is the duty of the society to enrich thereby enlighten its citizens. Only then the educated audience will appreciate the speaker properly.

If the culture is rich then the audience musters strong in the poetry corner beside the street to listen the poems of the new poets. But, if the people are uneducated then they are addicted to enjoy the dance program of the bar coupled with wine of night club casino for violent enjoyment.

There are two types of persons. The first category takes credit through speaking. The second category enriches him through listening. Both are diagonally opposite. The speaker is so blunt that hardly can he realize none laughs for him, rather everybody laughs at him.

Listening is an attribute. The teacher advises listen carefully. This attribute has to be acquired from childhood. A good listener prospers always than a speaker. A wise can listen the voice of inner soul or conscience. He decides accordingly. A fool cannot hear it.

All cannot listen. One has to acquire it. It is an art. All is not artist. This answers why we observe few listeners around us. In fact, good listener nowadays, has become a fossil.

Someone likes to listen. He avoids speaking. To him speaking demands knowledge and intellect as well. It is a laborious job to speak logically. Listening enriches knowledge without any effort. It is cent percent gain. Listening is not to bear all and everything. One must protest but after full listening for full gain. One who speaks more lies more and forgets most. In contrast one who hears more gains more.

Lacking in listening is a chronic problem. If a person does not stop or rectify himself then everybody is speaker. There is no listener. It is a decorum that one has to keep mum for the sake of patience hearing when someone speaks. Someone communities are famous for speaker. They have gift of the gab. Some communities are famous for listener. They are come by nature.

Through study a person becomes giant of knowledge. Also listening renders a person inventory of knowledge. Knowledge is power. An intelligent person gathers knowledge and thereby enriches himself from both the avenues.

There is good speaker. But there is scarcity of good listener. Everybody tries to highlight his name thereby acquire fame. Everybody is busy to beat his own drum. None is ready to hear but to speak. In fact listening is a great virtue. All cannot listen. It demands perseverance. Perseverance is akin to physical pain. Very few people can bear it. This answers why we see few good listeners around us.

Good listening helps to take proper and fruitful action. Some persons begin to quarrel without full hearing. This half hearing is dangerous more than no hearing. If a person takes action without hearing then there is cause of no hearing. But a partial hearer cannot claim of non-hearing. Thus he is penalized for incomplete hearing.

CONCLUSION

In case of love both the lovers will speak and listen to each other. Both the eloquent lovers are made for each other. But an eloquent and silent partner is mad for each other for being opposite in characteristic nature and behaviour. Both these categories are good lovers. Now, if the partners are both silent then they are great lovers. They say where goodness ends, greatness begins. Silence is more eloquent than speaking anything. Silence beckons. It implies consent. Both enjoy unfathomable voice of listening through eyes.

REFERENCES

No reference, since the present article is an outcome of Creative Nonfiction Writing.

OF CRITICISM

ABSTRACT

There are two types of persons. The first type likes to criticise and the second type likes to be criticised. Both the persons do their job untiringly. None can restrict them. No force can resist them. The first type criticises always; whether the matter is good or bad it matters little. His job is to criticise. As such he criticises always. The second type bears the torture of the critic silently and does his duty sincerely. He seldom minds and never declines to do his job. He is like a shock absorber or a lighting arrester. He is really a Good Samaritan. Thus he is a benefactor of mankind.

KEYWORDS: Criticism, comment, review, fault finding, disapproval, remark, judgement

INTRODUCTION

Creative writing is based more on manifestation rather than on expression. It does not inform, rather it reveals. So it bears no reference. The best creative writing is critical, and the best critical writing is creative. This article is an outcome of thinking about creative writing meant for a general readership. As such, I have adopted a free style methodology so that everyone can enjoy the pleasure of reading. As you might know, Francis Bacon (1561-1626), the immortal essayist, wrote many essays namely 'Of Love', 'Of Friendship', 'Of Ambition', 'Of Studies', and so on. The multiple-minded genius correctly pointed out that all the words of the dictionary can be used as themes for essays. But little has been done since his death to continue or finish his monumental task. Bacon's unique individual style of presentation ignited my imagination and encouraged me to write creative essays as a method of relieving a wide range of emotions through catharsis.

ARTICLE

Criticism is the act, art, or principles of criticising, especially of criticising literary or artistic work. It is a comment, review, article, etc. expressing this. It implies faultfinding. It is disapproval. It is the art of judging, especially in literature or the fine arts. It is a critical judgement or observation. It is the work of a critic. It is critical article, essay or remark.

Criticism is the action or process of indicating the faults of somebody or something or one's disapproval of somebody or something. It is a remark that indicates a fault or faults e.g. in response to the criticism levelled at senior officers. It is the art of making judgements on literature, art, etc. e.g. literary criticism. It is such a judgement.

Criticism is a negative approach. Its outcome is always negative. Everybody hates it. Everybody is afraid of it. But none can avoid it. Someone can bear it. Someone cannot at all. Thus it is a derogatory event. Animadversion and review are synonym of criticism.

Criticism is the act or art of analysing and judging the quality of something, especially a literary or artistic work, musical performance, dramatic production, etc. It is the act of passing severe judgement. It is to censure. It is a critical comment, article, or essay. It is critique. It is any of various methods of studying texts or documents for the purpose of dating or reconstructing them, evaluating their authenticity, etc.

Criticism is a statement or remark expressing this. It is the work of a critic. It is an article, essay, etc., expressing or containing an analytical evaluation of something. The higher criticism is the criticism dealing with the origin and character etc. of texts, especially of biblical writings. The lower criticism is the textual criticism of the Bible. Higher or historical criticism, as distinguished from textual or verbal criticism, is the inquiry into the composition, date, and authenticity of the books of the Bible, from historical and literary considerations. Textual criticism deals with and seeks in correct reading of especially MS text of an author.

Criticism is the act of making judgements. It is analysis of qualities and evaluation of comparative worth. It is especially the critical consideration and judgement of literary or artistic work. It is a comment, review, article, etc. expressing such analysis and judgement. It is the art, principles, or methods of a critic or critics. It is the scientific or scholarly investigation of texts or documents to discover their origin, history, or original form.

Man criticises. He has to criticise. He is bound to criticise. Similarly he is criticised. He faces criticism. He is destined to face criticism. In all these events he is quite undone. Thus man, willy-nilly, faces criticism infinite times from cradle to coffin. Someone criticises to save himself or to serve and satisfy his egoistic attitude or both simultaneously.

It is good to criticise. It is better not to criticise. It is best to avoid it. As such a fool criticises. A wise avoids it tactfully. Thus tact of omission is a great virtue. All cannot avoid it. Avoidance is an art. All is not artist. This answers why most of the people criticise and be the cause of criticism as well.

One should not lose heart for destructive criticism or be reckless for constructive criticism. In case of former one he must rectify himself and in case of later one he must follow the correct avenue always. As such a wise welcomes criticism be it constructive or a destructive one.

Criticism is of various types. Also it differs in different degrees and different dimensions as well.

A scheme that is open to criticism becomes effective when precaution measures are taken before implementation. This practise is seen in private firm but seldom in the public sector. The bank's actions attract heavy, strong and widespread criticism. A fickle-minded person hates and can't take criticism i.e. being criticized. Planning is thinking before doing. As such a wise invites criticisms before any planning for its smooth implementation. It is done for the safe voyage. The learned knows that in case of failure everybody and even a fool will criticise him.

Critic is one who pronounces judgement. He is censurer. He is judge of literary or artistic works, especially professional reviewer of books etc. He is one skilled in estimating the quality of literary or artistic work. He is a professional reviewer.

Critic is a person who judges, evaluates, or criticises. He is a person who customarily, as for his occupation, judges the qualities or merits of some class of things, especially of literary or artistic works, dramatic or musical performances, etc. He is a person who tends too readily to make captions, trivial, or harsh judgements. He is a person who expresses a low opinion of somebody or something.

Critic is a person who forms and expresses judgements of people or things according to certain standards or values. He is such a person whose profession is to write or broadcast such judgements of books, music, paintings, sculpture, plays, motion pictures, television, etc., as for a newspaper. He is a person who indulges in faultfinding. He is one skilled in textual studies, various readings, and the ascertainment of the original words. Thus he is a person engaged in textual criticism.

Reviewer, judge, connoisseur are synonyms of critic.

There are various subjects. As such there are various types of critics depending on concerned subject or subject concerned. There is music critic. There is theatre critic. There is literary critic. Criticism is a tough job. All cannot be a critic. It needs talent, good temperament, stamina and perseverance as well. Constant practise sharpens the act of criticism. The outcome of criticism depends on the gradation of the intellect of the critic. Since criticism is qualitative in nature, there may be different criticism on a single topic.

A play much praised by the critic experiences financial gain. A wise is his own severest critic. She confounded her critics by breaking the record i.e., they said she would not be able to do so, but she did.

Every body's mother is no body's mother. As such public project, generally, fails, since there is none to look after it. A critic avails himself of this opportunity. He takes credit forecasting the failure of the public project. Thus he is acknowledged as wise and experienced. It is a win without any battle. He gets promotion without any toil but applying his talent. He enjoys sadistic pleasure.

Criticaster is an inferior, minor or incompetent critic. He is a petty critic. Criticism is an honourable job, but for this type of callous man both critic and criticism get hatred.

Critical indicates the faults in somebody or something or one's disapproval of somebody or something e.g. a critical remark or report. The inquiry was critical of her handling of the affair. It is derogatory in nature. For example: Why are you always so critical?

Critical is an attribute of or relating to the judgement or analysis of something, especially literature, art, etc. For example: In the current critical climate her work is not popular. The movie has received critical acclaim i.e. praise from the critics. Try to develop a more critical attitude, instead of accepting everything at face value.

Critical is of or at a crisis. For example: We are at a critical time in our country's history. The patient's condition is critical i.e. he is very ill and may die.

Critical is of the greatest importance. It implies crucial. For example: The depth of the foundations is critical. The growing undercurrent of public protest proved a critical factor. Speak critically of somebody. He is critically ill.

Critical means inclined to find fault or to judge with severity, often too readily. It is occupied with or skilled in criticism. It is involving skilful judgement as to truth, merit, etc. It is judicial e.g. a critical analysis. It is of or pertaining to critics or criticism e.g. critical essays. It is providing textual variants, proposed emendations, etc. e.g. a critical edition of Chaucer. It is pertaining to or of the nature of a crisis. It is crucial e.g. a critical moment. It is involving grave uncertainty, peril, etc. It is dangerous e.g. a critical wound.

In physics it is pertaining to a state, value, or quantity at which one or more properties of a substance or system undergo a change. In mathematics it indicating a point on a curve at which a transition or change takes place.

Critical is a making or involving adverse or censorious comments or judgements. It is expressing or involving criticism. It is skilful at or engaged in criticism. It is providing textual criticism e.g., a critical edition of Milton. It is of or at a crisis. It is involving risk or suspense e.g. in a critical condition, a critical operation. It is decisive, crucial e.g. arrive at the critical moment, matter of critical importance.

Critical is characterised by careful analysis and judgement e.g. a sound critical estimate of the problem. It is of critics or criticism. It is of or forming a crisis or turning point. It is dangerous or risky. It is causing anxiety e.g. a critical situation in international relations. It is of the crisis of a disease. It is designating or of important products or raw materials subject to increased production and restricted distribution under strict control, as in wartime. It is designating or of a point at which a change in character, property, or condition is effected or at which a nuclear chain reaction becomes self-sustaining.

Critical, faultfinding, captious, cavilling, carping are synonymous.

Critical, in its strictest use, implies an attempt at objective judging so as to determine both merits and faults e.g. a critical review. But it often and hypocritical, always connotes emphasis on the faults or shortcomings. Faultfinding implies a habitual or unreasonable emphasis on faults or defects. Captious suggests a characteristic tendency to find fault with, or argue about, even the pettiest details e.g. a captious critic. Cavilling stresses the raising of quibbling objections on the most trivial points e.g. a cavilling grammarian. Carping suggests peevishness, perversity, or censoriousness in seeking out faults.

Critical is at or relating to a turning point, transition or crisis. It implies seriously ill. It is relating to criticism. It is rigorously discriminating. Critical is censorious, fault-finding e.g. was critical of me, my efforts. It is skilful at or engaged in criticism. It is providing textual criticism e.g. a critical edition of Ibsen. It is belonging to criticism. Critical is pertaining to a crisis, involving risk or suspense e.g. critical condition, operation.

There is a good critic of drama, a dramatic critic, a very poor sort of art critic. Bradley is one of the great Shakespeare critics. Practising philanthropy one will find a host of critics at his/her heels. There are critical opinions at variance with poetic practice. One should be free from critical habits for the sake of his own benefit. Critical moment or critical condition causes anxiety.

Carping, faultfinding, cavilling, discriminating, nice, exact, precise, decisive, climacteric, hazardous, precarious, perilous are synonym to critical.

Antonym to critical is unimportant.

There are critical apparatus, critical angle, critical mass, critical point, critical path, and critical temperature having unique meaning respectively. Critical philosophy is that of Kant which is based on a critical examination of the faculty of knowledge.

Criticise is to form and express a judgement on a work of art, literature, etc. It is teaching students how to criticize poetry. It is to make judgements as to merits and faults. It is to discuss critically. It is to pass judgements on. It is to analyze and judge as a critic. It is to judge disapprovingly. It is to censure. Criticise is to indicate the faults of somebody or something. For example: Stop criticizing my work. He was criticized by the committee for failing to report the accident. He criticized by taking unnecessary risks.

Criticise, reprehend, blame, censure, condemn, denounce, cavil, appraise are synonymous.

Criticise, in this comparison, is the general term for finding fault with or disapproving of a person or thing. Reprehend suggests sharp or severe disapproval, generally of faults, errors, etc. rather than of persons. Blame stresses the fixing of responsibility for an error, fault, etc. Censure implies the expression of severe criticism or disapproval by a person in authority or in a position to pass judgement. Condemn and denounce both imply an emphatic it pronouncement of blame or guilt. Condemn suggesting the rendering of a judicial decision, and denounce public accusation against persons or their acts. Cavil is to make unnecessary complaints about something. It is to criticise for the sake of criticism to make situation more complicated. Appraise is to assess the value, quality or nature of somebody or something. It is positive in nature.

Antonym of criticise is praise.

Critique is a critical analysis. For example: The book presents a critique of the government's economic policies. Critique is an article or essay criticising a literary, architectural, or some other work. It is review. It is a criticism or critical comment on some problem, subject, etc. It is an instance or the process of formal criticism. Critique is a critical essay analysis or evaluation of a subject, situation, literary work, etc. It is the act or art of criticising. It is criticism. Critique is a critical examination of any production. It is a review. It is to discuss critically.

There are two types of persons. The first type likes to criticise and the second type likes to be criticised. Both the persons do their job untiringly. None can restrict them. No force can resist them. The first type criticises always whether the matter is good or bad it matters little. His job is to criticise. As such he criticises always. The second type bears the torture of the critic silently and does his duty sincerely. He seldom minds and never declines to do his job. He is as if, a shock absorber like, a lighting arrester. He is really a Good Samaritan. Thus he is a benefactor of mankind.

Further critics are of two types viz. good critic and bad critic. The good critic criticises and highlights the merit. Also he points out the demerit of the matter in question quite sympathetically. He pleads for constructive criticism. He knows the toil required for any creation. He knows where the shoe pinches. But the second type of critic is so blind that he cannot and can never see the merit of anything. He always finds fault with anything thus presented to him for his valued opinion. His criticism is destructive always. He enjoys sadistic pleasure hurting the creator. It seems unshaded infant coupled with unguarded childhood offered him such a unique personality pattern. He suffers from severe criticism. No body laughs for him. Everybody laughs at him. Yet he does not change his nature. In fact, personality pattern cannot be changed at that belated period of life. He is so cursed. Criticism dogs him wherever he goes. He faces criticism till he breathes his last.

Again there is another type of critic who, as per his sweet will, criticises and opines what the creator did not mean at all. He expresses and imposes his personal innovative opinion upon the creator. He considers it as his democratic right. The most interesting part of such democracy is that the creator did not dream at all of such explanation in his dream even. Such a critic may be called super critic. If the manuscript is discovered after the death of the creator then the interpretation of super critic breaks all barriers and crosses all boundaries with violent emotion. Since the creator is dead then so many critics appear with so many interpretations on a single topic. No force can bind them. None can restrict them.

CONCLUSION

They say, he who can does and he who cannot criticises. Since a critic has no creative power, he cannot create anything. As such he does suffer from inferiority complex. This very complex renders him timid and causes fear from writing as well like hydrophobia. This lacking in capacity provokes him to be envious of the creator. He hates a creator and attacks him through destructive criticism thus to compensate his deficiency thereby to serve and satisfy his egoistic attitude. He is afraid of any kind of writing. He seldom writes anything lest his writing should be criticised. He is merciless in criticism which is the manifestation and outcome of violent anger. He knows well that his rubbish manuscript will also be severely criticised and returned back as boomerang. He will be paid back by his own coin. Thus the critic himself is singularly liable to create his own Frankenstein.

REFERENCES

No reference, since the present article is an outcome of Creative Nonfiction Writing.

OF INTELLECT

ABSTRACT

If the husband is dull-headed then he is well-advised to take guidance, in each and every step, from his intellectual wife to gain both ways. In the first situation for any wrong decision, by his wife, he cannot be held responsible. But in case of success he will enjoy benefit, being merely a sleeping partner, without having any intellect.

KEYWORDS: Intellect, reasoning, perceive, power, thinking, mind, rational, meaning

INTRODUCTION

Creative writing is based more on manifestation rather than on expression. It does not inform, rather it reveals. So it bears no reference. The best creative writing is critical, and the best critical writing is creative. This article is an outcome of thinking about creative writing meant for a general readership. As such, I have adopted a free style methodology so that everyone can enjoy the pleasure of reading. As you might know, Francis Bacon (1561-1626), the immortal essayist, wrote many essays namely 'Of Love', 'Of Friendship', 'Of Ambition', 'Of Studies', and so on. The multiple-minded genius correctly pointed out that all the words of the dictionary can be used as themes for essays. But little has been done since his death to continue or finish his monumental task. Bacon's unique individual style of presentation ignited my imagination and encouraged me to write creative essays as a method of relieving a wide range of emotions through catharsis.

ARTICLE

Intellect is the ability to reason or understand or to perceive relationships, differences, etc. It is the power of thought. It is faculty of knowing and reasoning. It is the mind, in reference to its rational powers. It is great mental ability. It is the thinking principle. It implies meaning. It is mental powers. It is high intelligence. It is a mind or intelligence, especially a superior one. It implies understanding.

Intellect is the power or faculty of the mind by which one knows or understands, as distinguished from that by which one feels and that by which one wills. It is the faculty of thinking and acquiring knowledge. It is capacity for thinking and acquiring knowledge. It is a particular mind or intelligence, especially of a high order.

Intellect is the faculty of reasoning, knowing, and thinking, as distinct from feeling. It is the power of the mind to think in a logical manner and acquire knowledge e.g. a woman with a keen intellect and exceptional qualities of leadership. It is the understanding or mental powers of a particular person etc. e.g. his intellect is not great. It implies a clever or knowledgeable person. It is the intelligentsia regarded collectively e.g. the combined intellect of four universities.

Intellect is a person of high intelligence. It is person, persons collectively, of good understanding. It is a person of great intelligence and powers of reasoning e.g. She was one of the most formidable intellects of her time. It is a person possessing a great capacity for thought and knowledge. It is minds or intelligent persons, collectively.

Reason, sense, brains are synonymous to intellect.

Intellection is the process of using the intellect. It is the exercise of the intellect. It is a particular act of the intellect e.g. The theory of Relativity is one of the sublime acts of intellection of all time. It implies thinking. It is cognition. It is an act of the intellect. It is a thought or perception. It is the action or process of understanding, especially as opposed to imagination. It is the act of understanding. As per philosophy, it is apprehension or perception.

Knowledge is power. Similarly, intellect is all and everything. It is a crucial factor for rising and lack of it causes utter and untimely downfall. It determines difference between two persons. With it a man enjoys win and without it he experiences defeat. As such it is the most important personality trait of any character. Man, from cradle to grave, experiences intelligence in its different forms and features along with diversified degrees and dimensions having various faces and facets as well.

A person may suffer from superiority complex when he finds others around him are inferior to him regarding intellect, wealth, etc. Similarly, in the reverse situation that very person suffers from inferiority complex. Complex of any kind is not good at all. As such, a wise seldom does suffer either from superiority or inferiority.

Animals are devoid of intellect. For this factor they are within the grip of human being. Elephant has less brain in comparison with its robust physique. Due to this lacking in reasoning power animals like dinosaur do not exist anymore. Intellect is the prerogative of human beings. A man of intellect as contrasted with an emotional man can be isolated at ease. The outstanding intellect of the age paved the way for huge development. A civilised nation is recognised by the intellect of its society. All his life he was engaged in intellectual pursuits. He has an intellectual appearance.

There is difference between intellect and experience. An intellectual person may lack in practical sense, etiquette, etc. He may be less social. Yet he is valued much for his prudence that compensates all of his deficiencies. It is better to have intelligent foe than a friend who is fool. They say it is better to go to hell with intelligent persons rather to dwell in the heaven with fools. A person without intellect loses all belongings. A penniless person with intellect may earn much. So intellect is more powerful than wealth. For intellect itself is alias and akin to wealth. Money brings money. Similarly, intellect brings money as well. Here return is without investment of money. So, intellect is equivalent to money and even more than that. As time passes man gathers experience. An intelligent person utilises previous experience and be more successful. A dull-headed person cannot use or utilise experience gathered so far for mundane gain. Here intelligence creates difference.

Intellect, generally, is inherited. A person gets it by birth. Regular practice sharpens it. It provokes a person to think. Thinking is akin to physical pain. Very few persons can bear that pain. This answers why we see very few intellectual persons around us.

If an intellectual person is rich then he can multiply his money very easily. Such moneyed persons are numbered. Yet they rule the world. In democracy majority rules the government. But in case of capitalist set up it is just reverse. Here minority rules the proletariat who are major in population.

Some persons have high intellect. Some persons have low intellect. Some persons are rich. Some persons are poor. Thus on the basis of these four distinct personality traits the whole population can broadly be classified into four different categories.

The first category is high intellect but poor. Such a person may become rich in course of time through his prudence. Obviously, luck is a great factor. Because, all highly qualified medical practitioners or lawyers may not have lucrative income. In contrast an ordinary doctor enjoys much popularity thereby earns much. However, real talent definitely shines in life today or tomorrow.

The second category is low intellect but rich. This richness may be due to inheritance or getting a lottery. But such a person can neither multiply his money nor can he hold the existing asset due to lack of practical judgement.

The third category is both low intellect and poor as well. Such a person loses more and becomes poorer. They are common mass and large in number. They are less educated and working class. Such human resource is the headache of the society thereby nation at large.

The fourth category is both high intellect and rich. Such persons are asset and build the nation. They multiply their money within short period of time and become richer. If such a person is kind-hearted then he earns popularity and wins the election easily. And if such a talented ruler becomes tyrant then he becomes either dangerously brilliant or brilliantly dangerous or both simultaneously. Then the plight of the public is beggar description.

Again, an intellectual person may be educated or uneducated i.e. may have no formal education at all. He may either be bold or timid. He may be a criminal. In case of low talent he commits petty crime. In case of high talent he becomes a finished scoundrel and leaves no clue at all of his crime thus committed.

An intellectual person may be emotional. Then he is tender in nature. He may be cultured. In such a case the society becomes culturally enlightened.

If an intellect is both beautiful and smart then he/she becomes hero/heroine. Such a glamorous heroine becomes the heart throb of young boys. An intellect ugly person may be the hero of his concerned field. A beauty queen with dull head may propose an intellect but ugly person. She thinks that her baby will inherit her face and brain of her husband thereby will be the topper in future. The paradox is that inheritance does not obey any rule. Then if the baby gets the face of its father and brain of its mother then it will be the worst outcome of this unpredictable venture.

An intellect may either be proud or live with a low profile. The latter one knows that pride goes before a fall. Again an intellect may be reserve or liberal personality. The person may also be extrovert or introvert. In case of extrovert they say he who speaks more lies more.

An intellect person may be elite or ordinary. An intellect hailing from elite class is always ahead by birth. They appear in this world to rule. Again, an intellect may come from all walks of life. He may be the embodiment of the down-trodden people. In that case the goal is far more than the elite intellect.

The eyes of intellect always spark. Beauty attracts seldom sparks. A dull-headed beautiful face is beautiful till it speaks not. An ugly person draws attention for its talent. A king is honoured in his own domain only, but a wise is worshipped everywhere in every ages.

An intellect person may be energetic or idle. In case of idle he becomes dangerous since he stocks his intellect. They say a non-practising lawyer is dangerous more than a practising lawyer. A practising lawyer gets the chance to discharge his intellect and gets relief through Catharsis. But the non-practising lawyer becomes a marginal player and waits for the prey.

An intellect may love. The drawback of an intellect is that he does not laugh. If an intellect practises to laugh then he can become an international lover. Laugh is all and everything whether intellect or dull. A smile can conquer the head and heart of a tyrant even. A fool laughs always spontaneously. Laugh offer intense heavenly delight. A fool gets it. A rich or an intellect is deprived from this natural blessings. An intellect dies due to heart attack for having excessive facts and figures. He is a giant of knowledge.

Someone laughs for someone. Someone laughs at someone. Only an intellect can judge the hidden reason. A fool laughs thrice. At first, he laughs seeing others to laugh. He laughs second time realising the reason of laughter. He laughs third time remembering that he laughed without knowing the reason. Thus he enjoys series of laughter which is quite enviable trait to someone who cannot laugh or have no lover for reciprocation.

There is a debatable question who can enjoy the world more an intellect or a fool.

The merit of a fool is that it can easily be moulded. The demerit of a fool is that it can easily be moulded simultaneously. In contrast, an intellect has only merit. He has no demerit. As such the merit of an intellect is that he cannot easily be moulded. Here lies the uniqueness of intellect and triumph of intellect as well.

CONCLUSION

If the husband is dull-headed then he is well-advised to take guidance, in each and every step, from his intellectual wife to gain both ways. In the first situation for any wrong decision, by his wife, he cannot be held responsible. But in case of success he will enjoy benefit, being mere a sleeping partner, without having any intellect.

REFERENCES

No reference, since the present article is an outcome of Creative Nonfiction Writing.

OF ASSUMPTION

ABSTRACT

Someone assumes depending on full logic. He is full genius. Someone assumes without any logic. He is non genius. Someone assumes with half logic. He is half genius. Someone assumes from partial knowledge. He is partial genius. Thus non genius is better than incomplete genius in reality. These variations in assumptions are the manifestation of various perfections already in public from all walks of life.

KEYWORDS: Assumption, assume, conception, hypothesis, supposition

INTRODUCTION

Creative writing is based more on manifestation rather than on expression. It does not inform, rather it reveals. So it bears no reference. The best creative writing is critical, and the best critical writing is creative. This article is an outcome of thinking about creative writing meant for a general readership. As such, I have adopted a free style methodology so that everyone can enjoy the pleasure of reading. As you might know, Francis Bacon (1561-1626), the immortal essayist, wrote many essays namely 'Of Love', 'Of Friendship', 'Of Ambition', 'Of Studies', and so on. The multiple-minded genius correctly pointed out that all the words of the dictionary can be used as themes for essays. But little has been done since his death to continue or finish his monumental task. Bacon's unique individual style of presentation ignited my imagination and encouraged me to write creative essays as a method of relieving a wide range of emotions through catharsis.

ARTICLE

Assumption is a thing that is thought to be true or certain to happen, but is not proved e.g. an implicit/underlying assumption. What leads you to make that assumption? The theory is based on a series of false/wrong assumptions. We are working on the assumption that the rate of inflation will not increase next year. They made certain assumptions about the market.

Assumption is the act of taking for granted or supposing. It is something taken for granted. It is a supposition e.g., a correct assumption. It is act of taking to or upon oneself.

Assumption is the act or an instance of assuming. It is presumption. It is impudence. It is reception. It is arrogance. It is the minor premise in a syllogism in logic.

Thus assume is to take or accept as being true, without proof, for the purpose of argument or action e.g. assume as a working hypothesis. It is a thing assumed in this way.

Assumption is a supposed bodily ascent into heaven. It is the taking up of the body and soul of the Virgin Mary into heaven after her death. Assumption, theist, in Christian thought, the doctrine that the Virgin Mary was 'assumed' i.e. taken up and received bodily into heaven. It dating from the 4th century is a doctrine held by the Roman Catholic and Orthodox Churches on the 1st November 1950. The feast in honour of this is held on 15 August as declared a dogma.

Assumptionist is a member of the Roman Catholic congregation i.e., Augustinians of the Assumption founded at Nimes in 1843.

Deed of assumption, as per Scots law, is a deed executed by trustees under a trust-deed assuming a new trustee or settlement.

Further assumption is the action of taking on power or responsibility. It is an act of taking possession of something e.g. the assumption of power; assumption of a charge/responsibility; his haughty assumption.

Assumption of something is an act of taking or beginning to have power, authority, etc. e.g. her assumption of supreme power; the assumption of an active role in regional settlements.

Synonyms of assumptions are conjecture, guess and forwardness.

Also assume, pretend, feign, affect, simulate are synonymous.

Assume implies the putting on of a false appearance but suggests a harmless or excusable motive. It is to seize. It is to usurp e.g. an assumed air of bravado; an assumed innocence/cheerfulness; an assumed name; an assumed engineer's status. It is to seize.

Pretend and feign both imply a profession or display of what is false, the more literary feign sometimes suggesting an elaborately contrived situation e.g. to pretend not to hear; to feign deafness.

To affect is to make a show of being, having, using, wearing, etc., usually for effect e.g. to affect a British accent.

Simulate emphasizes the imitation of typical signs involved in assuming an appearance or characteristic not one's own e.g. to simulate interest.

Man assumes to proceed. It is an optimistic venture. It implies positivism already in man. It is the manifestation of activeness. A lazy person seldom assumes. He likes to stand still. He is afraid of movement for the uncertainty pregnant with injustice and fear as well. An emotional person assumes if he likes. He assumes not if he likes not. Thus his mood and motive are gloriously so uncertain.

Man assumes. He has to assume. He is bound to assume. Thus man willy-nilly assumes infinite times from cradle to coffin. His assumptions vary in different forms and features having different degrees and dimensions as well depending on the concerned case or case corned as is faced with.

He assumed office/assumed control of the organization this morning. His illness assumed a grave character. Well, let us assume that for the sake of argument. He assumed to himself all credit for the success. His scholarly assumptions are intolerable.

Someone assumes something considering it appropriate. He adopts measure accordingly. Someone assumes to move forward. Logical assumption paves the way to reach the desired goal. Illogical assumption or baseless assumption not only wastes time and money rather it causes irritation to all concerned. In case of crisis period risky assumption either pays or perishes. Only a judicious brain can tackle this problem properly.

Man assumes since he does not know the answer. He assumes to reach the goal. Assumption of fool is wrong for lacking in reasoning. Wise assumes correctly. From assumption and subsequent success there from one can gauze the intellect of the concerned person. Similarly, failure caused by wrong assumption confirms the dullness of the fool.

Scientist assumes god as merely a hypothesis. Pious soul believes in the existence of god. The sacred heart visualizes god through hallucination thereby enjoys intense heavenly delight. He enjoys sanctity. He enjoys mental peace and happiness. In contrast an atheist explains hallucination as merely an optical illusion. Thus belief is the essence of theist. Disbelief is the capital of atheist. Both are diagonally opposite in philosophy towards their life.

A wise seldom takes action from assumption. He decides upon confirmation of the fact. A fool takes action basing on assumption without verifying the genuineness of the rumour. It is the serious drawback of the fool. But the paradox is that without assumption one cannot reach the goal. Here logical assumption is the solution. As such assumption is the business of wise or logician never of a fool.

A person finds that his assumptions sometimes are correct. Sometimes he realises that his assumptions are incorrect. Thus practical experiences enrich him to attain perfection.

A person needs not to assume when he finds the desired thing readymade. Here luck favours him to achieve the desired thing. Correctness of assumption depends on the intellect. Intellect varies person to person. Guess varies accordingly. So is the success.

A person assumes and gets relief. He remains restless till he assumes. He becomes mentally free after any kind of assumption. It pains him much to remain in vacant mood. As such he is eager to be engaged with any assumption right or wrong.

Someone assumes depending on full logic. He is full genius. Someone assumes without any logic. He is non genius. Someone assumes with half logic. He is half genius. Someone assumes from partial knowledge. He is partial genius. Thus non genius is better than incomplete genius in reality. These variations in assumptions are the manifestation of various perfections already in public from all walks of life.

A fool is rigid and is reluctant to change. A judicious brain assumes, proceeds and changes decision accordingly whenever change is required to find the desired result. In this way he gathers experience from trial and error method the most important tool of any scientific investigation. This type of investigation is noted for its thoroughness. It has no gap or loopholes.

CONCLUSION

Assumption is a conception. Wrong assumption misguides and invites danger. Someone corrects. Someone corrects not. The second category, being a fool, suffers and loses always. He does not know how to rectify. Here lies his limitation. Wrong assumption is not rare. It is quite common. Even a judicious brain becomes the victim of wrong notion. The wise believes in correctibility and decides accordingly forthwith as and when required. That's why he is great. It is detrimental only when false vanity or silly ego or baseless anger or proud foolishness or altogether closes the door for taking correct assumption.

REFERENCES

No reference, since the present article is an outcome of Creative Nonfiction Writing.

OF PARADOX

ABSTRACT

A successful politician is one who can conquer the head and heart of the public at ease. The paradox is that that very successful politician fails successfully to befool his wife. His wife knows very well his real identity.

KEYWORDS: Paradox, contrary, self-contradictory, strange, opposite, absurd, unlikely

INTRODUCTION

Creative writing is based more on manifestation rather than on expression. It does not inform, rather it reveals. So it bears no reference. The best creative writing is critical, and the best critical writing is creative. This article is an outcome of thinking about creative writing meant for a general readership. As such, I have adopted a free style methodology so that everyone can enjoy the pleasure of reading. As you might know, Francis Bacon (1561-1626), the immortal essayist, wrote many essays namely 'Of Love', 'Of Friendship', 'Of Ambition', 'Of Studies', and so on. The multiple-minded genius correctly pointed out that all the words of the dictionary can be used as themes for essays. But little has been done since his death to continue or finish his monumental task. Bacon's unique individual style of presentation ignited my imagination and encouraged me to write creative essays as a method of relieving a wide range of emotions through catharsis.

ARTICLE

Paradox is a person, thing or situation that has two contrary features and is therefore rather strange. It is a statement containing opposite ideas that make it seem absurd or unlikely although it is or may be true. For example: 'More haste, less speed' is a well-known paradox. It is the use of this in writing, etc. e.g. a work full of paradox and ambiguity.

Paradox is a statement or proposition seemingly self-contradictory or absurd but in reality expressing a possible truth. It is any person, thing or situation exhibiting an apparently contradictory nature. It is an opinion or statement contrary to commonly accepted opinion.

Paradox is a statement contrary to common belief. It is a statement that seems contradictory, unbelievable, or much absurd but that may be true in fact. For example: "Water, water, everywhere, nor any drop to drink". It is a statement that is self-contradictory and, hence, false. It is a person, situation, act, etc. that seems to have inconsistent qualities or full of contradictions.

Paradox is seemingly absurd though perhaps actually well-founded statement e.g. hydrostatic paradox. It is a person or thing conflicting with a preconceived notion of what is reasonable or possible. It is a paradoxical quality or character. It is a statement contrary to accepted opinion.

Paradoxical is of, having the nature of, or expressing a paradox or paradoxes. It is of or like or involving paradox. It is fond of using paradoxes. It is seemingly full of contradictions.

Paradox ides is a typically Middle Cambrian genus or trilobites, some very large about two feet (60cm) long.

Paradoxology is the utterance or maintaining of paradoxes.

Paradoxy is the quality of being paradoxical.

Paradoxical sleep is apparently deep sleep, with actual increased brain activity.

Paradoxure is a civet-like carnivore of Southern Asia and Malaysia, the palm-cat of India. It is a palm civet, animal of genus Paradoxurus with remarkably long curving tail.

Malfunctioning of brain offers paradoxical statements which are inconsistent in nature. Such an insane character loses link of his talks thereby talks rubbish. In fact, in gossip paradoxical statement amuses the audience much. A wise never speaks paradox. Even, if he says inadvertently, he admits it instant on interrogation or identification. As such it is enjoyable more to speak with a foe having full of logic rather than a friend who talks illogically.

In science there is no paradoxical result, rather the methodology thereby observation is incorrect. There may be two identical twins. There may be two identical results. But two identical wrong results are impossible. Truth is one. Lies are many. It means result differs due to different methodologies thus adopted. When the methodology is identical result also is identical.

A man may have dual character but seldom a non-living being. This is due to the fact that man is rational and thing is irrational. Rationality has its both merit and demerit simultaneously. In case of wise rationality is merit. In case of fool or evil it is demerit. Because, a dishonest person does not do what he commits. He does not confess what he did.

Virus acts like both living and nonliving being. It is not paradox. Sometimes it acts like a living being. Sometimes it behaves like nonliving creature. It possesses two characters. Frog is amphibious. It can move both in land and water. Metal may either be in solid or in liquid state. Iron is solid. Mercury is liquid. But iron or mercury cannot remain in different state as found in normal temperature.

A word is not ambiguous by itself, rather it is used ambiguously. Ambiguity is not paradox. Thus all paradox is ambiguous but all ambiguous may not be paradox.

There is no paradox, but paradoxical or confusing policy. There are different methodologies or tactics to cheat or befool a person thus to gain the game. There is no paradox, but use thing paradoxically.

Regarding any person if it is stated that (1). The person is cheater, (2). The person is honest. This is an example of pure paradoxical statement. A person can never simultaneously be both cheater and honest. He has a unique identity. Either he is cheater or honest. Obviously, a finished scoundrel can pretend honestly.

Man speaks paradox and faces paradox infinite times from cradle to grave in indifferent forms and features with different faces and facets having various degrees and dimensions as well.

He always created controversies by uttering paradoxes. The hydrostatic paradoxes confuse a learner at ease. As a figure of speech, paradoxes are quite common. How the best team could lose the match was a paradox.

A notorious person speaks paradox for gain and if not then applies force. Profit is his single agenda. He intends to attain it by hook or by crook or by any means fair or foul. Only a judicious can take right decision. A dishonest speaks lie i.e. paradox. Lie is alias and akin to paradox. Thus plurality of lies gives birth to paradox. Paradox is always ambiguous. One paradox gives birth to another paradox.

She sings well but cannot speak is a paradox. He who cannot speak can never sing. Ugly heroine is a paradox. A heroine can never be ugly. Unsmart hero is a paradox. Hero is always smart. Both heroine and hero must have glamour. A callous can't be a hero.

A mathematician may be dishonest. It is not paradox. Dishonesty can never be a bar to be a mathematician. Angry hero is not a paradox. Hero may either be angry or cool. These are merely personality traits. Here, appearance i.e. glamour is the base factor in film industry. For his baseless anger all the heroines may avoid him. He may remain bachelor. Bachelor hero or bachelor heroine is not a paradox. Common people like to see their hero and heroine to lead a happy conjugal life. But the paradox is that the romantic couple of film may avoid each other in reality. Paradoxically, some heroines love rough type hero.

A father must have child. It is not paradox. That child may be born before or after marriage. Marriage is not the bar to give birth to a child. In a conservative society there is strict sequences i.e. marriage, sex and then issue. In such an orthodox culture, free sex or dating is prohibited. Thus love child is illegal for the sake of the chastity of the society. But desertion of wife and children are illegal and more than immoral.

It is a paradox that professional comedians often have unhappy lives. Professional life and personal life are two diagonally opposite things. Someone masters on profession. Someone masters on personal. Only a fortunate soul masters on both. A shrewd talented person can earn money thereby prosper in business successfully. But that very shrewd may fail successfully to earn family peace. That is a third factor and can only be achieved by a blessed soul.

A successful politician, generally, is unhappy with family life. Similarly, a religious leader is unhappy with his mad wife. In fact, happiness binds and unhappiness releases a person from home to go to the outer world. Unhappy family life of both paves the way for outward success. Both of them become compelled to leave family. Then the whole world is their family. Thus they are of the people, by the people, and for the people. The paradox is that their earnings are looted by the people. In this way they achieve greatness.

Now, those who want to enjoy happy family life they are well advised not to join politics. Politics means polite or tricks or poly tricks. All cannot follow and track this meaning. Similarly, a happy person should not be ambitious to be a religious God-father. A greedy person wants to please both wife and public simultaneously. He can please none. Those who want to please everybody can please no body. He loses both wife and public. Then he has no where to go. He becomes a homeless vagabond.

In the light of professional and personal life human being can broadly be classified into four different categories.

The first category is professionally happy but personally unhappy. They earn money. Spend money for wife. But the paradox is that their wives are most unhappy with their rich husbands for unknown reasons or for the reasons better known to their better half. In fact for happiness money is a must. But, the paradox is that, money alone is not all and cannot guarantee for full happiness accordingly.

The second category is professionally unhappy but personally happy. They are idle. They are satisfied with little income. Their wives have less demand. The wife of such a person loves her husband very dearly, since the obedient husband always stays nearly to serve and satisfies her needs clearly. The person compensates his financial deficiency through physical labour. Even a rich person is envious of his divine happiness. Thus, money cannot purchase divinity.

The third category is both professionally and personally happy. They are either made for each other or mad for each other or both simultaneously. Such a rich person is a blessed soul and ideal as well. But the paradox is that ideal is always is unattainable to the common mass.

The fourth category is both professionally and personally unhappy. They are mad. They have to beg to maintain their livelihood. They commit suicide, since they do not know that life has its sweet side too.

Incomplete knowledge gives birth to paradox. Incomplete knowledge is alias and akin to half knowledge. Then such a half genius cannot speak logically due to lack of knowledge. Rationality helps to detect wrong and necessary correction accordingly. A paradox if found instant can be corrected instant. Long gap or isolated event cannot prove easily the paradox faced in the past dawn.

CONCLUSION

A successful politician is one who can conquer the head and heart of the public at ease. The paradox is that that very successful politician fails successfully to befool his wife. His wife knows very well his real identity.

REFERENCES

No reference, since the present article is an outcome of Creative Nonfiction Writing.

OF ARISTOCRACY

ABSTRACT

Rome was not built in a day. Similar is the aristocracy. It cannot be purchased. It has to acquire. It needs time. It demands time. It is not instant. It is a slow process. It is not a blackbody that absorbs heat immediate and loses heat immediate. A tree is known by its fruit. A man is known by his behaviour. Similar is the case of an aristocrat person. He is noted for his pleasing personality. He is always positive. Thus he is optimist. Wise is the real aristocrat. A king is aristocrat in his own domain only, but wise is worshipped everywhere. Aristotle is aristocrat. Here lies the uniqueness of aristocracy.

KEYWORDS: Aristocracy, highest class, noble birth, nobility, privileged class, blue blood

INTRODUCTION

Creative writing is based more on manifestation rather than on expression. It does not inform, rather it reveals. So it bears no reference. The best creative writing is critical, and the best critical writing is creative. This article is an outcome of thinking about creative writing meant for a general readership. As such, I have adopted a free style methodology so that everyone can enjoy the pleasure of reading. As you might know, Francis Bacon (1561-1626), the immortal essayist, wrote many essays namely 'Of Love', 'Of Friendship', 'Of Ambition', 'Of Studies', and so on. The multiple-minded genius correctly pointed out that all the words of the dictionary can be used as themes for essays. But little has been done since his death to continue or finish his monumental task. Bacon's unique individual style of presentation ignited my imagination and encouraged me to write creative essays as a method of relieving a wide range of emotions through catharsis.

ARTICLE

Aristocracy is the highest class in society. It implies nobility. It is the nobility as a ruling class. It is government by or political power of a privileged order. It is a state governed in this way. It is the best representatives or upper echelons. It is a class of persons holding exceptional rank and privileges, especially the hereditary nobility.

Aristocracy is the government by the best or most outstanding citizens. It is a state so governed. It is ruling body of nobles. It is class of nobles. It is the best representatives of intellect, etc. It is a government by the best or most able men. It is a governing body composed of the best or most able men in a state. It is any class or group considered to be superior. It is government by a privileged minority or upper class, usually of inherited wealth and social position. It is a country with this form of government. It is oligarchy. It is a privileged ruling class. It is an analogous class in respect of any quality.

Thus aristocracy means people of noble birth or rank i.e. members of the aristocracy. An aristocracy of talent implies the most able or talented members of a society. It is a government or state ruled by an aristocracy, elite, or privileged class. It is those considered the best in some way e.g., aristocracy of scientists. It implies aristocratic quality or spirit.

Aristocracy of intellect implies outstanding scholar. Labourer gets minimum wages. Aristocracy of labour implies higher wages. Then, such a labourer is no more a member of the working class. But he cannot be considered as the elite class for his less talent or low education or both simultaneously. Now he becomes an envious person of an elite member whose income is lesser than him.

Aristocrat is a member of an aristocracy. He is a nobleman. He is a person of noble birth. He is a person with the tastes, manners, beliefs, etc. of the upper class. He is a person who believes in aristocracy and an advocate of an aristocratic form of government. Also aristocrat is a haughty person.

Aristocratic is of, characteristic of or favouring aristocracy as a form of government. It is of an aristocracy or upper class. It is like or characteristic of an aristocrat. It is having the manners, values, or qualities of the aristocracy e.g. aristocratic bearing. It is used in either a favourable sense i.e. proud, distinguished, etc. or an unfavourable i.e. snobbish, haughty, etc. It is grand. It is stylish. Thus aristocratic name, aristocratic family, aristocratic manner, aristocratic lifestyle, etc. represent blessed people.

Rome was not built in a day. Similar is the aristocracy. It cannot be purchased. It has to acquire. It needs time. It demands time. It is not instant. It is a slow process. It is not a black body that absorbs heat immediate and loses heat immediate. A tree is known by its fruit. A man is known by his behaviour. Similar is the case of an aristocrat person. He is noted for his pleasing personality. He is always positive. Thus he is optimist.

Aristocratic life is very costly. To maintain aristocracy it needs huge money. An aristocrat person has blue blood. But common people exist with RBC i.e. red blood corpuscles. Elite is alias and akin to aristocrat. Behaviour of elite sharply differs with a common man. He can easily be isolated for his personality. A person, of high birth, if misbehaves is ill-famed. He loses his status. He is no longer aristocrat at all. On the contrary, if a person hailing from low birth if behaves with politeness, he attains the status of an aristocrat personality.

Goodness is alias and akin to aristocracy. It is difficult to acquire aristocracy. It is difficult more to hold it for long. An aristocrat woman marries another man who is aristocrat by birth. But a non-aristocrat man if wants to marry an aristocrat woman, he must either be very rich or highly educated or both simultaneously. Here complexion or handsome appearance accelerates to conquer the heart of the elite woman.

Someone opts for aristocracy. Someone opts for money while selects life-partner. The equation differs from person to person. Here not voice rather choice dominates. A judicious person selects a rising rich groom, for his daughter, than a poor guy who was once elite. This choice is opted so that his daughter may live with solvency. But that very person selects a bride of aristocrat family even the family is no more rich. This decision is just to enrich his culture by the presence of the bride hailing from aristocrat family.

Money is a must. But it is not all. All cannot be purchased with money. It has limitation. Mere money seldom offers aristocracy. Again aristocracy without money is quite absurd. To maintain aristocracy money is highly required. Nothing exists in the vacuum. Money enriched with culture is the yardstick of aristocracy.

Behaviour is the chief ingredient of aristocracy. It is the most precious. Real aristocrat is learned and well-behaved. He is known by behaviour. He is rich in knowledge. A man may lose wealth, but he does not lose aristocracy which is manifested from behaviour or appearance. Similarly aristocracy is not a commodity. Aristocrat is rich. A person may have money but may not be aristocrat. Thus all aristocrats are rich, but all rich may not be aristocrat. Richness in behaviour confirms real noble birth.

All are not aristocrat. It is of some, by some and for some privileged persons. As such aristocracy of talent and aristocracy of intellect are very rare. Aristocrats are numbered. They are minor. Common people are major. Generally, majority rules the minor. But, the paradox is that, aristocrat people rule the society. It is just reverse. Thus minority rules majority. Privileged class always rules and becomes more privileged and proletariat becomes poorer. It is so-called aristocracy preferred and practised by the bourgeois. The communist agitates against it.

CONCLUSION
Wise is the real aristocrat. A king is aristocrat in his own domain only, but wise is worshipped everywhere. Aristotle is aristocrat. Here lies the uniqueness of aristocracy.

REFERENCES
No reference, since the present article is an outcome of Creative Nonfiction Writing.

OF ABSTRACT

ABSTRACT

Since abstract is not concrete, it suffers from plurality. Also it is enriched thereby pregnant with plural ideas. Abstract is alias and akin to obscure. It is ambiguous as well simultaneously. It dwells at the threshold between knowledge and beyond knowledge. From any abstract one may guess something of the whole that may or may not be correct. It is the store house of vague or baseless ideas. It reveals less suppresses more. As such abstract misguides more than to guide. It is tip of the iceberg. It is the tiny part of infinity.

KEYWORDS: Abstract, thought, idea, abstruse, theory, art, extract, essence, summary

INTRODUCTION

Creative writing is based more on manifestation rather than on expression. It does not inform, rather it reveals. So it bears no reference. The best creative writing is critical, and the best critical writing is creative. This article is an outcome of thinking about creative writing meant for a general readership. As such, I have adopted a free style methodology so that everyone can enjoy the pleasure of reading. As you might know, Francis Bacon (1561-1626), the immortal essayist, wrote many essays namely 'Of Love', 'Of Friendship', 'Of Ambition', 'Of Studies', and so on. The multiple-minded genius correctly pointed out that all the words of the dictionary can be used as themes for essays. But little has been done since his death to continue or finish his monumental task. Bacon's unique individual style of presentation ignited my imagination and encouraged me to write creative essays as a method of relieving a wide range of emotions through catharsis.

ARTICLE

Abstract means existing in thought or as an idea but not having a physical or practical existence. It is general. It is not based on any particular person, situation, etc. e.g. talk about something in abstract terms/in an abstract way. In art it is not representing people or objects in a realistic way but expressing the artist's ideas and feelings about certain aspects of them e.g. an abstract painting/painter/design/ballet.

Abstract is an abstract idea or quality. It is an example of abstract art e.g. a painter of abstracts. It is a short summary of a book, etc. e.g. an abstract of a lecture. The abstract is ideal or theoretical way of regarding things. It is idealistic, not practical. It is abstruse. In art etc. it is free from representational qualities.

Abstract is conceived apart from any concrete realities, specific object, or actual instance e.g. an abstract idea. It is expressing a quality or characteristic apart from any specific object or instance. It is theoretical. It is not practical or applied e.g. abstract science. It is abstruse e.g. abstract speculations. In fine arts it is of or pertaining to the formal aspect of art, emphasizing lines, colours, generalized or geometrical forms, etc., especially with reference to their relationship to one another. It is pertaining to the nonrepresentational art styles of the 20th century. It is a summary of a statement, document, speech, etc. It implies epitome. It is something that concentrates in itself the essential qualities of anything more extensive or more general, or of several things. It implies essence. It is an idea or term considered apart from some material basis or object. It is an abstract work of art.

Abstract is thought of apart from any particular instances or material objects. It is not concrete. It is expressing a quality thought of apart from any particular or material object e.g. beauty is an abstract word. It is not easy to understand. It is theoretical. It is not practical or applied. It is designating or of art abstracted from reality, in which designs or forms may be definite and geometric or fluid and amorphous e.g. a generic term that encompasses various non-realistic contemporary schools. It is a brief statement of the essential thoughts or content of a book, article, speech, court record, etc. It is an abstract thing, condition, idea, etc. It is to take away. It is to remove quietly. It is to take dishonestly. It is to steal. It is to think of a quality apart from any particular instance or material object that has it. It is also, to form a general idea from particular instances. It is to summarize. It is to make an abstract of.

Abstract is to do with or existing in thought rather than matter, or in theory rather than practice. It is not tangible or concrete e.g. abstract questions rarely concerned him. It is of a word especially a noun e.g. denoting a quality or condition or intangible thing rather than a concrete object. It is not aiming to depict a representation of external reality. It is often followed by from to take out of. It is to extract. It is to do this as an occupation. It is often followed by from to disengage a person's attention etc. It is to distract. It is followed by from to consider abstractly or separately from something else. It is an abstraction or abstract term.

Abstract is to draw away. It is to separate. It is to purloin. It is to separate by the operation of the mind, as in forming a general concept from consideration of particular instances.

Abstract is a summary, abridgement e.g. Shakespeare, Antonio and Cleopatra III, vi., explained by some as an abridgement of time of separation. It is others conjecture abstract. It is that which represents the essence.

Abstract is a thought of apart from any particular instances or material objects. It is not easy to understand. It is designating or of art that seeks to make an effect through form and colour alone. It is an abstract thing, condition, etc. It is to think of a quality apart from any particular instance or from any object that has it.

Abstract is separated from matter, practice, or particular examples. It is to deduct. It is to euphemise. It is to consider abstractly. Justice in the abstract is nothing.

It is the abstract means something that exists only as an idea e.g. the abstract versus the concrete. It is to draw or take away. It is to divert or draw away the attention of. It is to consider as a general quality or characteristic apart from specific objects or instances e.g. to abstract the notions of time, space, and matter.

In the abstract is an idiom. It means in a general way, without reference to a particular person, thing, event, situation, etc. For example: Consider the problem in the abstract. It is without reference to practical considerations or applications. It is in theory e.g. beauty in the abstract. It is an abstraction. It is in theory as distinct from practice.

Abstract something from something is to remove something. It is to separate something from something else. For example: Two other points must be abstracted from the argument. It is to make a written summary of a book, etc.

Abstracted is thinking of other things. It implies not paying attention e.g. an abstracted gaze/smile. It is lost in thought. It is deeply engrossed or preoccupied. It is removed or separated from something. It is withdrawn in mind. It is absent-minded. It is inattentive to the matter in hand.

Abstracted is apart from actual material instances, existing only as a mental concept. It is opposite to concrete e.g. away from practice. It is theoretical of terms denoting a quality of a thing apart from the thing, as redness. It is representing ideas in geometric and other designs, not the forms of nature i.e. paint and culture. It is drawn off with form. It is withdrawn in thought. It is not attending.

Abstraction is an abstract idea e.g. ideological abstraction. It is the quality of being abstract. It is the state of thinking of other things and not paying attention. It is absent-mindedness e.g. a general air of abstraction. It is the action of removing something from something else. It is the state of being removed from something else e.g. ideas conceived in abstractions from physical observations.

Abstraction is an abstract or general idea or term. It is an idea that cannot lead to any practical result. It is something visionary and unrealistic. It is the act of considering something as a general quality or characteristic, apart from any concrete realities, specific object, or actual instance. It is the act of taking away or separating. It is the state of being lost in thought. In, Fine Arts, it is the abstract qualities or characteristics of a work of art. It is a nonrepresentational work of art.

Abstraction is an abstracting or being abstracted. It implies removal. It is formation of an idea, as of the qualities or properties of a thing, by mental separation from particular instances or material objects. It is an idea so formed, or a word or term for it e.g. honesty and whiteness are abstractions. It is an unrealistic or impractical notion. It is mental withdrawal. It is absent-mindedness. It is an abstract quality. It is a picture, sculpture, statue, etc. that is wholly or partly abstract.

Abstraction is the act or an instance of abstracting or taking away. It is an abstract or visionary idea. It is the formation of abstract ideas. It is abstract qualities especially in art. It is an abstract work of art.

Abstraction is taking away, withdrawal. It is euphuism i.e. stealing. It is process of stripping an idea of its concrete accomplishments. It is an idea so stripped, something visionary. It is piece of abstract art. It is absence of mind.

Abstraction is the act of abstracting. It is the state of being abstracted. It is abstract quality or character. It is withdrawal from worldly things. It is absence of mind. It is a purloining. It is the process of abstracting by the mind. It is a thing existing only in idea. It is a theory, visionary notion. It is an abstract term. It is an abstract composition of paint and culture.

Abstraction is the process of stripping an idea of its material accompaniment. Democracy in our country has been reduced to an abstraction. Abstraction is necessary for the classification of things. It is a visionary idea.

Abstraction is formation of an idea, as of the qualities of a thing, by separating it mentally from any particular instances or material objects. It is an idea so formed, or a word for it e.g. honesty is an abstraction. It is preoccupation.

Abstractionism, in Fine Arts, is the theory and practice of abstract art. It is the principles and practice of abstract art. It is the pursuit or cult of abstract ideas.

Abstractionist is a person who paints abstract paintings. He deals in abstractions or unrealities. It is of or pertaining to abstract art.

Abstract noun, in English grammar, is a noun having an abstract or general rather than a concrete or specific referent, as dread or greyness. It is a noun that refers to an abstract quality or state, e.g. goodness or freedom. It is not a physical object. It denotes quality or state.

Abstract algebra is the branch of mathematics that deals with the existence of algebraic concepts usually associated with the real number system to other, more general systems.

Abstract expressionism, in Fine Arts, is a style of painting in which paint is applied in an apparently random manner producing images that may or may not have reference to forms exterior to the picture.

Abstract expressionism is a post- World War II movement in painting characterized by emphasis on the artist's spontaneous and self-expressive application of paint in creating a nonrepresentational composition.

Abstract expressionism is a movement in American painting associated with New York in the 1940s and 1950s, whose subscribers laid particular emphasis on the creative spontaneous act and the role of the unconscious and unfettered emotion in the process of painting. Strongly influenced by surrealism, and most often associated with the spontaneous gestural style of action painting and figures such as Pollock and de Kooning, the term also refers to the vast 'all-over' canvass of colour-field painting associated with Mark Rothko, Barnet Newman, and Clyfford Still (1904-80).

Abstract expressionism is a style of painting popular after World War II, in which the artist's self-expression is carried out by applying paint freely in compositions that do not represent known objects.

Abstract expressionism is action painting.

Abstractive is having the power of abstracting. It is pertaining to an abstract or summary. It is that abstracts or can abstract. It is of or having to do with abstraction. It is able or tending to abstract. It is formed by or pertaining to abstraction. It is anything abstractive. It is an abstract.

Abstract of title, in Law, is an outline history of the title to a parcel of real estate, showing the original grant, subsequent conveyances, mortgages, etc. It is a brief history of the ownership of a piece of real estate, from the original grant through the present holder, including a statement of liens to which it may be subject. Thus it is a summary of facts concerning ownership.

Abstract art is in its broadest sense, any art that does not attempt to represent external, recognizable reality, and which thereby tends to concentrate on shape, colour, and form. Specifically, the term relates to a modern phenomenon, the first abstract works of art are generally considered to date from 1910-14, in which the traditional Western view of art imitating nature was abandoned, and which is manifested in a great variety of movements and styles. From these, two general types can be distinguished e.g. hard-edged and geometric akin to the linear and classical, and flowing and organic akin to the painterly and romantic.

Abstracter or abstractor is one who makes abstracts with or for a grade of Civil Service clerks.

Abstract is of the scholar, by the scholar and for the scholar. It is the business of the scholars only. It is quiet obscure to a fool. As such a fool has nothing to do in this regard. In fact it is beyond capacity of the fool.

Abstract has its various faces and facets. Also man faces it in different forms and features having varied degrees and dimensions as well.

Abstract science and abstract art both demand and depend on high intellect of brain. His speech was too abstract. A good lucid abstract of the speech draws attention and lasts long in the memory of the reader. Abstract is extract of metal from ore. It is abstract the notion of dimension from that of space. It is quietly abstracting the chain from the girl's neck. We may talk of beautiful things, but beauty itself is abstract. He has some abstract notion of wanting to change the world.

Some scholars write abstract in a lucid manner to make one understand. Some others write in obscure style to highlight talent. Thus the difficult texts remain unread and gather centurion dusts in the corner of library. It may be merit of a scholar but demerit of the readers. They say he who realises anything clearly can also explain it lucidly. So lucidness of any abstract thereby the manuscript as a whole is the yardstick to gauze the intellect of the concerned scholar. Thus obscurity of any manuscript doubts the talent of the concerned scholar rather confirming it instead.

Since abstract is not concrete, it suffers from plurality. Also it is enriched thereby pregnant with plural ideas.

Abstract is alias and akin to obscure. It is ambiguous as well simultaneously. It dwells at the threshold between knowledge and beyond knowledge. From any abstract one may guess something of the whole that may or may not be correct. It is the store house of vague or baseless ideas. It reveals less suppresses more. As such abstract misguides more than to guide. It is tip of the iceberg. It is the tiny part of infinity.

Life and death are two great abstract things or thinkings diagonally opposite to each other. Life means mundane existence. Death means similar existence elsewhere beyond perception. Here lies the place of abstract thinking. Rebirth is a never ending abstract thinking to the philosophers and theologist since time immemorial. It is an unsolved abstract idea and most debatable as well.

Abstract allows someone to think independently without any restriction. It is quite democratic in nature and behaviour. To some serious persons mundane existence is merely abstract thing. They like to think seriously. They hate casual people and idea as well. They want to enjoy mundane life through abstract thinking rather than mundane. They avoid casual thinking and people as well.

CONCLUSION

Abstract is a subject matter of wise seldom of fool. The fool is for mundane, the wise is for divine. Since mass is more light of abstract reaches to few only. To the mass abstract remains as grey area for ever. On the other hand the fortunate few souls can illuminate this dark assertion. To the rest, abstract remains as un-illuminated.

REFERENCES

No reference, since the present article is an outcome of Creative Nonfiction Writing.

OF ABUSE

ABSTRACT

Abuse has its own grammar. This grammar is not known to all. A person may know it but does not use it for the sake of status. A Good Samaritan avoids it. Those who care not for status use abusive language. In fact dirty language or dirty culture degrades an innocent mind. Then the society hates and avoids such a diverted person who becomes isolated from the main stream.

KEYWORDS: Abuse, wrong, improper, misuse, harmful, injurious, offensive, deceive, cheat

INTRODUCTION

Creative writing is based more on manifestation rather than on expression. It does not inform, rather it reveals. So it bears no reference. The best creative writing is critical, and the best critical writing is creative. This article is an outcome of thinking about creative writing meant for a general readership. As such, I have adopted a free style methodology so that everyone can enjoy the pleasure of reading. As you might know, Francis Bacon (1561-1626), the immortal essayist, wrote many essays namely 'Of Love', 'Of Friendship', 'Of Ambition', 'Of Studies', and so on. The multiple-minded genius correctly pointed out that all the words of the dictionary can be used as themes for essays. But little has been done since his death to continue or finish his m onumental task. Bacon's unique individual style of presentation ignited my imagination and encouraged me to write creative essays as a method of relieving a wide range of emotions through catharsis.

ARTICLE

Abuse is to use wrongly or improperly. It is to misuse e.g. to abuse one's authority; to abuse a privilege; abused his position of power. It is to treat in a harmful, injurious, or offensive way e.g., to abuse one's eyesight. It is to deceive. It is to cheat. It is to masturbate. It is to use harshly, coarsely, insulting or bad language. It is bad or improper treatment. It is mistreatment. It is a bad, unjust, corrupt, improper practice or custom e.g., the abuses of bad government.

Abuse is to betray as confidence. It is to revile. It is to violate. It is to hurt by treating badly. It is injury. It is an evil, unjust or corrupt practice. It is deceit. It is undue advantage. It is betrayal of confidence. It is ill usage. It is deception. It is misrepresentation.

Abuse is to use to bad effect or for a bad purpose. It is to take a drug for a purpose other than a therapeutic one. It is to be addicted to a substance. It is incorrect or improper use e.g. the abuse of power. It is maltreatment or especially sexual assault of a person. It is perversion. It is insulting or unkind speech.

Abuse is to make wrong, bad or excessive use of one's power, position etc. e.g. abusing drugs/solvent/alcohol; widespread abuse of computer facilities; an abuse of trust/privilege/authority. Further example: The new system of paying cash bonuses is open to abuse. It is to take undue advantage of one's power, position, etc. unfairly or excessively e.g. abuse somebody's hospitality; abuse one's authority/the confidence placed in one. It is to treat a person or an animal badly or violently e.g. a hostel for abused and battered women.

Abuse is to assault especially sexually. It is to have sex with somebody illegally or against their will e.g. viciously beaten and sexually abused; a man who abused his own daughter. It is to speak insultingly, harshly, unjustly or in an offensive way to or about somebody. For example: Journalists had been threatened and abused. He abused the confidence we placed in him.

Abuse is an unfair or illegal practice e.g. put a stop to political abuses; abuses of human rights. It is cruel treatment of a person or animal, especially sexually e.g. child sex abuse; physical abuse of horses. These are rude and offensive remarks about a person. These are insults e.g. verbal abuse; hurl i.e. a stream/torrent of abuse at somebody. Further example: The word 'bastard' is often used as a term of abuse.

Abusive is wrong. It is containing, giving of the nature of abuse. It is coarsely reviling. It is catachrestical. It is, of speech or a person, offensive and insulting. It is criticizing harshly and rudely e.g. abusive language/remarks. He became abusive.

Abuser is a person who abuses e.g., a drug abuser; a child abuser.

Misapply, wrong, mistreat, maltreat, injure, harm, hurt, vilify, berate, slander, defame, aspersion are synonymous to abuse.

Abuse, censure, invective all mean strongly expressed disapproval.

Abuse implies an outburst of harsh and scathing words against another often one who is defenceless e.g. abuse directed against an opponent; abused his power by profiting at the expense of others; abuse of power/privilege/confidence; put an end to long-standing abuses.

Censure implies blame, adverse criticism, or hostile condemnation e.g. severe censure of acts showing bad judgment.

Invective applies to strong but formal denunciation in speech or print, often in the public interest e.g. invective against graft.

Praise, compliment, eulogize, laud are antonym of abuse.

Abusage is wrong use, especially of words or grammar. Here the author, for defense, contends it as poetic license. The experienced author breaks the rule and sometimes crosses the boundary of grammar intentionally just to draw attention. But a novice author is well advised not to do this before earning reputation, otherwise it will be highlighted as mere ignorance which is a hindrance to gain fame.

Abuser, abusing, misuse, deception, wrong, outrage, reviling are synonymous.

Both fool and uneducated person abuse. Education acts as brake. As such an educated person seldom abuses. Yet if a so-called educated person abuses then it may be concluded that the person was not properly educated. Then such a person is equal and at par, regarding status, with an uneducated person.

Slang words are the vehicle of abuse or abuse is the vehicle of slang words. Some persons abuse always whether the topic is good or bad it matters little. He cannot express without slang words. When a person becomes the victim of violent emotion he loses emotional brake. Then he abuses. It is true even in case of so-called educated person. Then the person uses filthy language. Uneducated person abuses and gets relief. It is the easiest avenue to get mentally relief. If an educated person or a renowned person abuses it becomes news. Then numerous comments come back as a boomerang to the abuser.

A fool abuses for instant gain. It is a risky game. It cannot be an avenue to gain profit always. Rather the opponent will take revenge and sometimes more severely if chance favours. If a wise abuses then he becomes a fool for his foolish behaviour. Then he is no longer respected.

Abuse is quite natural and basic instinct. Education saves a person from such dirty affair. It polishes the personality. But in careless moment it outbursts and the person gets relief through catharsis. In using filthy languages both educated and uneducated persons are equal and at par. In culture and art excessive use of any dialogue or any dark colour is called abusing. Repetition i.e. nagging is abuse and causes irritation. Everybody avoids such a nagger.

Man enjoys sadistic pleasure through abusing. A fool enjoys sadistic pleasure using dirty words. The degraded souls knows better that a gentle man will not exchange slang words and readily leave the battle or surrender to it and give whatever it will demand as is happens in case of dacoit.

Abuse is a nasty game. It is the tool used by a perverted soul. Everybody avoids it for its negative and evil spirit. But all cannot avoid it. Everybody becomes the prey of it to a less and greater degree many times during entire life span.

Abuse has its own grammar. This grammar is not known to all. A person may know it but does not use it for the sake of status. A Good Samaritan avoids it. Those who care not for status use abusive language. In fact dirty language or dirty culture degrades an innocent mind. Then the society hates and avoids such a diverted person who becomes isolated from the main stream.

Abusive language should not apply on educated or good persons. Then it may react and come back as boomerang. A fool or a corrupted soul works only when instructions are given through abusive languages. They do not understand the language of enlightened society. But it may not always be correct.

Sweet behaviour or smile of beautiful face acts as good tonic. It seems there is no hard and fast rule of abusing. Always bad behaviour may not work well. As such, cocktail of both good and bad behaviour work well. Only an experienced expert can apply it properly to solve the problems as are faced with. Yet a wise always thinks for future consequences and never use filthy language.

CONCLUSION

In politics abuse is the nastiest tool used mercilessly and recklessly by the political parties to defame the opponent. But the paradox is that the defame returns back as boomerang. Common people do not care for these baseless allegations of the rival opponents. Only that political party wins in the election that can read the pulse of the voter.

REFERENCES

No reference, since the present article is an outcome of Creative Nonfiction Writing.

OF LENIENCY

ABSTRACT

Leniency proves the solvency of mind. It is an orientation. It highlights the broadness of mind. It is a positive philosophy towards life. A lenient person possesses pleasing personality. For this personality trait he is popular. Thus a lenient person is popular. Conversely, a popular person is lenient. If a person is not popular then he is not lenient at all.

KEYWORDS: Leniency, mercy, tolerant, easy going, forgiveness

INTRODUCTION

Creative writing is based more on manifestation rather than on expression. It does not inform, rather it reveals. So it bears no reference. The best creative writing is critical, and the best critical writing is creative. This article is an outcome of thinking about creative writing meant for a general readership. As such, I have adopted a free style methodology so that everyone can enjoy the pleasure of reading. As you might know, Francis Bacon (1561-1626), the immortal essayist, wrote many essays namely 'Of Love', 'Of Friendship', 'Of Ambition', 'Of Studies', and so on. The multiple-minded genius correctly pointed out that all the words of the dictionary can be used as themes for essays. But little has been done since his death to continue or finish his monumental task. Bacon's unique individual style of presentation ignited my imagination and encouraged me to write creative essays as a method of relieving a wide range of emotions through catharsis.

ARTICLE

Leniency is the fact or quality of being more merciful or tolerant than expected. It is clemency. For example: The court could show leniency.

Lenience is a concept. This concept is especially used in the context of punishments, like prison sentences. Thus it is treatment in which someone is punished or judged less strongly or severely than would be expected. For example: The defending lawyer asked for leniency on the grounds of her client's youth. But for the leniency of the referee, the player would have been sent off.

All accused does not get the favour of leniency. It depends upon many criteria. Only a fortunate accused gets the blessings of leniency. It is a difficult job of the judge to show such discretionary power. It creates much psychological pressure before passing any severe and serious verdict.

A lenient person is a far-sighted person. He is merciful. He is tolerant. Thus mercy and tolerance are two chief ingredients of a lenient soul. He is enriched with practical experience. The learned knows that mercy and tolerant have no substitute. They themselves are their substitutes. These are the investments of the learned for future return. Mercy pays. Revenge pays not. Tolerance offers much return even sometimes beyond expectation. Haste seldom pays rather it causes much loss.

A lenient person is permissive. He goes easy. He shows mercy which is above normal. Normal mercy is not lenient. It must be beyond normal. Thus degree of mercy is the yardstick of leniency. If leniency favours someone then it is detrimental to another one. Undue favour is called nepotism. If everybody gets equal favour then the question of nepotism does not arise.

A wise is lenient. A fool cannot be lenient. To be lenient one has to be judicious. A novice cannot be lenient. He who believes in correctibility is lenient. The judge, through his discretionary power, shows leniency to the first level criminal to rectify or correct himself so that he can lead a happy and peaceful life in future. A serious offender does not get the benefit of leniency.

A fool cannot realise the inner significance of forgiveness. The fool believes in revenge. He wants to pay back by his own coin. But if someone is forgiven by a powerful person then the offender commits no wrong later on. In contrast if he is punished severely then he may commit the crime repeatedly thereby becomes habitual offender. So a pain cannot be the substitute of another pain but pleasure can. Pleasure though tender in nature possesses immense power of rectification.

A miser is not lenient. A mean-minded person cannot be lenient. He cannot think of it in his dream even.

Leniency is an art. All are not artists. All cannot be artist. This answers why we observe few lenient persons around us.

Leniency proves the solvency of mind. It is an orientation. It highlights the broadness of mind. It is a positive philosophy towards life. A lenient person possesses pleasing personality. For this personality trait he is popular. Thus a lenient person is popular. Conversely, a popular person is lenient. If a person is not popular then he is not lenient at all.

A person of high birth is lenient. A person of low birth is not lenient. High birth helps to be lenient. Low birth restricts to be lenient. Thus culture acts as a catalyst to be lenient. Now, if a person of high birth cannot show leniency he is degraded as low birth. Conversely, if a person of low birth shows leniency then he attains the status of elite class. Thus a blessed soul is lenient. And a cursed soul is deprived from such divine blessings.

A person acquires leniency by birth. Good culture nurtures thereby paves the way for its full blooming. Also good culture renders a person lenient. Bad culture diminishes thereby destroys it. It is an added quality that isolates a person from others and keeps always ahead.

A lenient person is not rigid. A rigid person may be lenient. Someone is so hardcore that he is always rigid. Someone is so tender thereby holds that divine power always. Thus they are diagonally opposite with each other.

Man gives leniency. Also man takes leniency. Both this give and take are equal in number as are experienced from cradle to coffin.

A wise person is guided by head. A lenient person is guided by heart. Emotion is the driving force of a lenient person. He dreams for ideal state. He lives in hard reality but thinks for ideal. Ideal is always unattainable. As such he is deceived many times. Yet he does not lose hope. He is so positive. He contends that leniency is tender in nature yet it can convert a rigid character into a lenient soul.

Rigidity is good. Always rigidity is bad. Rigidity is good if it favours. Every favour confirms the disfavour of the opponent.

A lenient person is either a wise or a fool. A wise is lenient intentionally to enjoy pleasure. But a fool is lenient for lacking in intellect. He cannot imagine future consequences. Further there is a third category of person who becomes lenient under compulsion. Later on he withdraws leniency when chance favours. As such one should be cautious before taking any kind of lenient help.

A lay man considers a lenient person as a fool. A wise person can recognise another wise person. Similarly, a wise can disclose the true identity of a lenient person. Only a wise person can appreciate the philanthropy of such a Good Samaritan.

CONCLUSION

If a person cannot reciprocate or return he is well advised not to take any kind of lenient help from anybody. In this world nothing is free. Every action either demands or hopes for equal or more return with interest. If return is not received properly in time then problem arises. In fact leniency is not free.

REFERENCES

No reference, since the present article is an outcome of Creative Nonfiction Writing.

OF BLUNDER

ABSTRACT

A student commits mistake in the examination. A man commits blunder in life. Love is an involvement. Love is a mistake. Marriage is a greater involvement. Marriage is a blunder. A lover commits mistake in selecting life partner. A person commits blunder if marries. Thus where mistakes ends, blunder begins.

KEYWORDS: Blunder, mistake

INTRODUCTION

Creative writing is based more on manifestation rather than on expression. It does not inform, rather it reveals. So it bears no reference. The best creative writing is critical, and the best critical writing is creative. This article is an outcome of thinking about creative writing meant for a general readership. As such, I have adopted a free style methodology so that everyone can enjoy the pleasure of reading. As you might know, Francis Bacon (1561-1626), the immortal essayist, wrote many essays namely 'Of Love', 'Of Friendship', 'Of Ambition', 'Of Studies', and so on. The multiple-minded genius correctly pointed out that all the words of the dictionary can be used as themes for essays. But little has been done since his death to continue or finish his monumental task. Bacon's unique individual style of presentation ignited my imagination and encouraged me to write creative essays as a method of relieving a wide range of emotions through catharsis.

ARTICLE

Blunder is to make a gross or stupid mistake, especially, through carelessness or mental confusion. It is to move or act blindly, stupidly, or without direction or steady guidance. It is to act or speak clumsily. It is a foolish tactless remark. It is to stumble. It is to mismanage. It is to botch.

Mistake of higher degree is called blunder. Also if the mistake is serious in nature, it is also blunder. So to do a blunder higher mental capacity is a must. As such a wise commits a blunder whereas a fool commits a mistake. If the mistake is serious in nature then its future consequence is much. If the blunder is rectified immediately then the situation remains within control, otherwise the whole project will be frustrated and should be started afresh which involves both time and money.

Man makes blunder in hurried moment. If the guess is not correct then incorrect assumption confirms mistake. When brain does not work logically then man commits mistake. In case of blind attempt there is every possibility of committing mistake. Sometimes the situation becomes so uncertain that a person commits mistake. If someone waits and does not get the desired result then he commits mistake. And if someone leaves the place and the chance appears just after of his departure then it is also mistake. Here option determines fate. Only a lucky soul wins the game.

Lack of experience causes confusion that confirms blunder. As such a novice commits blunder. Here lies the importance of guide. Parents must guide their children lest they are not diverted. In case of girl issue mother must protect her and guide properly so that she is not sexually abused.

Everybody commits blunder. In this regard both wise and fool are equal and at par. The difference is a wise can identify and rectify the blunder which the fool cannot. At old age man loses memory. As such an old person commits blunder frequently.

If a situation goes out of control, due to a blunder, then only an expert can solve that problem. The situation becomes severe if one mistake is replaced by another mistake. As time passes situation becomes more serious. Similarly, the situation becomes grave due to committing series of mistakes. The objective goes far and remains unachieved.

There is silly mistake. But there is no silly blunder. Mistake may be silly. But silly is not attributed upon blunder. As such someone commits a silly mistake. None commits a silly blunder. In contrast, blunder enjoys higher status. There may be great blunder. There may be Himalayan blunder. He who commits great blunder is great. Similarly, he who commits Himalayan blunder is superior in characteristic nature and behaviour than others. These two types of personalities can easily be identified from their respective past records.

Also there is gross mistake. In office a clerk commits a mistake which is ill famed as clerical mistake. All liabilities or onus are imposed upon clerks. But there is no officer mistake. It would be better if clerical mistake would be termed as official mistake like technical mistake.

They say he, who does, makes a mistake. In contrast, he who does never, makes never any mistake. As such a shrewd does nothing thereby ill famed never. In contrast a fool rushes everywhere and mistakes everywhere. A Good Samaritan serves the ailing humanity. He dedicates his life for the sake of philanthropy. As such he seldom cares for mistake or blunder, since his interest is always for noble cause.

None wants to make a blunder. Nobody likes it. Everybody wants to avoid it. But none can escape from it. Someone makes fewer mistakes. Someone makes frequent mistake. Someone commits mistake never. He is a finished scoundrel. Someone can rectify himself. Someone wants not to rectify. He enjoys making blunder. His life is a series of blunders. His life is the episodes of different types of blunders.

Blunder is omnipresent. Man commits it either inadvertently or unknowingly. Thus man unfortunately commits blunder infinite times from cradle to coffin. In this regard he is quite undone except experiencing the sequences quite helplessly. Thus blunder appears with its immense presence, in its various forms and features having different degrees and dimensions as well. Problems caused by committing blunder at early period of life can somehow be managed, but that at late hours of life may not be tackled due to old age.

Everybody commits blunder. As such everybody is the prey of it. Blunder levels all. It is quite democratic in its nature and behaviour. As such to it, literate or illiterate, wise or fool, rich or poor, high birth or low birth, bourgeois or proletariat i.e. people from all walks of life are equal and at par.

In fact, man commits blunder inadvertently infinite times from cradle to coffin. Man commits blunder. He has to pay for it. Sometimes he can compensate the loss caused by blunder. Then he can move ahead. An unfortunate person cannot make up that loss. He suffers. His suffering depends upon the degree of blunder. A tender soul who is sentimental in nature may fail to face the problems caused by blunder and commits suicide. Those who commit suicide are well advised that life has its sweet side too.

They say even God can do any mistake, but Satan seldom does any mistake. A finished scoundrel never commits any mistake. A Satan falls in that category.

Childhood is the most precious time of life to build up the career. If someone wastes that important time then thousand moments of future can hardly compensate that deficiency. Man cannot realize it at that tender age. Later on when he realizes the blunder thus committed it is too late. He repents at that belated period of life. He mourns since he has reached the autumn of life. He curses himself. He fails everywhere. Misfortune dogs him wherever he goes. Sad luck follows him like shadow. He suffers till he breathes his last. Thus, if childhood is both unshaded and unguarded then sufferings are infinite and immense as well.

Blunder is ever companion of man like shadow. But it is not a friend at all. Rather it is a foe. It is solely responsible for the downfall or degradation of any soul. A mad person makes mistake. Also a vigilant person having sound body and sound mind makes mistake due to carelessness. As such carelessness or callousness is the root cause of making blunder. A lunatic person is permanent callous thereby permanent careless. So he is not entrusted with any work.

Machine makes no mistake. Man is not machine. So man commits mistake. Machine goes out of order when it works beyond stipulated period of time. Similarly, man becomes tired after prolonged work and becomes inattentive and makes mistake if works further in that wretched condition of both body and mind. So, rest is required to revitalize the tired nerves thereby engage at work again.

Someone commits bonafide mistake with malafide intention. Someone commits malafide mistake with bonafide intention. From the concerned mistake the motif of the concerned person can be ascertained. Here previous track record of the person in question illuminates the dark assertion.

A student commits mistake in the examination. A man commits blunder in life. Love is an involvement. Love is a mistake. Marriage is a greater involvement. Marriage is a blunder. A lover commits mistake in selecting life partner. A person commits blunder if marries. Thus where mistake ends, blunder begins.

Both love and marriage may or may not offer peace and happiness. A poor person may be happy. A rich person may be unhappy. A married employee may be happy. A married manager may be unhappy. He who cannot manage commits blunder. A tactless person commits blunder. A manager can manage his employee. He may not manage his wife. Through tact he can overcome any fault or blunder in his office. But he may fail successfully to implement that tact in his personal life.

In fact happiness is the outcome of good luck and unhappiness is the outcome of sad luck. Children may either be prize or punishment to the parents. None knows whether he will be rewarded or punished. Here mistake or blunder has no control over love or life.

They say bachelor lives like prince and dies like dog. An intelligent person does not like to love thereby wants not to make a mistake. A bachelor wants not to commit blunder inviting marriage. Emotion provokes a person to love and marry. Thus a brainless person loves and marries. A wise is guided by head and a fool is guided by heart.

Blunder compels a person to face awkward situation. It causes shame and ill-fame as well. If the blunder is due to inadvertence then it is quite common. If the blunder is intentional then it is illegal in the eye of law. Then the culprit is hated by all. Further, if the blunder is unsocial in nature then the accused is socially boycotted. This type of tyranny is observed in patriarchal society and women are its prime target since time immemorial.

CONCLUSION

Both love and marriage are risky games. A wise wants to love without taking any kind of risk which is next to impossible. As such a wise remains unloved or unmarried. Now a wise by mistake may love and by blunder may marry. The paradox is that, in reality, a fool is seen to lead a happy married life and a wise experiences painful married life. It seems, luck has an influential role on human life. None can control luck but luck controls everybody. Here blunder offers an extra dimension in its role. Here lies the uniqueness of blunder rather than unique blunder.

REFERENCES

No reference, since the present article is an outcome of Creative NonfictionWriting.

OF EASY

ABSTRACT

It may be easy to marry a beautiful woman but difficult to hold her for long, just like easy earning money is not always easy to keep in safe custody. In both the cases it needs stamina. It demands intellect. Thus those who do lack in stamina and intellect as well and are poor both in appearance and wealth are well advised neither to dream for beautiful wife nor should they earn easy money. Yet man runs after beautiful woman and easy money whether he can hold it or not it matters little. It is greed that provokes man to achieve the desired thing. Man enjoys holding instinct and boasts of for the same. Holding is owner's pride, neighbor's envy. Failure ignites the craze more violently. Craze is an easy avenue to enjoy the life to the lees at ease and easily.

KEYWORDS: Easy, difficult, vulnerable, informal, derogatory, careful

INTRODUCTION

Creative writing is based more on manifestation rather than on expression. It does not inform, rather it reveals. So it bears no reference. The best creative writing is critical, and the best critical writing is creative. This article is an outcome of thinking about creative writing meant for a general readership. As such, I have adopted a free style methodology so that everyone can enjoy the pleasure of reading. As you might know, Francis Bacon (1561-1626), the immortal essayist, wrote many essays namely 'Of Love', 'Of Friendship', 'Of Ambition', 'Of Studies', and so on. The multiple-minded genius correctly pointed out that all the words of the dictionary can be used as themes for essays. But little has been done since his death to continue or finish his monumental task. Bacon's unique individual style of presentation ignited my imagination and encouraged me to write creative essays as a method of relieving a wide range of emotions through catharsis.

ARTICLE

Easy means achieved without great effort. It implies presenting few difficulties e.g., an easy way

of retrieving information.

It is free from worries or problems e.g., promise of an easy life in the New World.

It is lacking anxiety or awkwardness. It is relaxed e.g., her easy and agreeable manner.

It is having no defence. It is vulnerable e.g., as a taxi driver he was an easy target.

It is informal. It is derogatory. It is very receptive to sexual advances, typically used of a woman.

For example: They thought she was easy, that they could buy her a drink and then get into her

pants at the end of the night.

It is without difficulty or effort. For example: We all scared real easy in those days.

It implies be careful. For example: Easy, girl—you'll knock me over!

Easy appears, in different phrases, with its various forms and features as follows.

Of easy virtue means very receptive to sexual advances, typically used of a woman. For example: Critics believed that as a painter she must be a woman of easy virtue.

Sleep easy is to go to sleep without worries. For example: Now you can sleep easy, safe in the knowledge that someone in a position of power is promoting your agenda.

Take it easy is to proceed in a calm and relaxed manner. For example: Well, in any case, you've got another few weeks around him, so I say take it easy.

Further take it easy means to make little effort. It is to take rest. For example: I decided to just relax and I took it easy all weekend long.'

Stand easy, in military, is used to instruct soldiers standing at ease that they may relax their attitude further.

Easy touch, informally, is a person who readily gives or does something if asked.

Easy does it is used to advise someone to approach a task carefully and slowly. For example: Easy, easy does it, not too much, just a little bit more.'

Easy come easy go is used, especially, in spoken English to indicate that a relationship or possession acquired without effort may be abandoned or lost without regret. For example: For him, allegations are easy come, easy go.

Easy on the ear, informally, means pleasant to listen to. For example: Her singing is easy on the ear.

I'm easy, informally, is said by someone when offered a choice to indicate that they have no particular preference. For example: If you don't want it to work, that's ok, I'm easy, I don't mind.

Easy on the eye, informally, means pleasant to look at. For example: A charming village that is easy on the eye.

Go easy on, informally, is to refrain from being harsh with or critical of (someone). For example: She's still nice and went easy on all the amateur performers.

Further, go easy on, informally is to be sparing in one's use or consumption of. For example: Club heroes watch what they eat, go easy on the drink and refrain from cigarettes.

Rest easy is untroubled by worries. For example: This insurance policy will let you rest easy.

Be easier said than done is to be more easily talked about than put into practice. For example: This is often easier said than done because it takes practice and commitment.

Take the easy way out is to extricate oneself from a difficult situation by choosing the simplest rather than the most honourable course of action. For example: She had taken the easy way out by returning the keys without a message.

Have it easy means have no difficulties. It implies be fortunate. For example: There is no single country that is having it easy.

It is easy to say anything, but difficult to do it. Most of the time starting is too easy and finishing is too difficult.

Nature is easy. It is easy for its unobstructed views. It attracts a romantic heart. City is not so easy. It is the outcome of destruction. After destruction of nature that easiness is lost forever. Nature soothes both eye and mind. It supplies fresh oxygen to the tired nerves that revitalize again. In contrast urban life suffocates thereby causes tension to lead a normal life.

One school of thought contends that there is conflict between right and easy. Another group declines its logic. A wise advises, "Do what is right, not what is easy". It is always beneficial to follow the right track that may take more time but output is unquestionable. Short cut method may be easy but it may cut short the span of life. Easy avenue if be illegal that it is not the avenue at all.

Easy access is both good and bad. One boss may be easy access. The second boss may not be so. He may punish the sub-ordinate for not maintaining formalities or protocol properly. As such they say look before you leap. The former is praised even after leaving. The later one is ill-famed permanently. If an inferior person asks to maintain the so-called formalities unnecessarily then he is laughed at. However, one must obey the rule and respect the chair. A boss enjoys all access. All is not boss. So all should not violate the rule thus to avoid punishment.

Knowledge manifests nobleness. Very few persons acquire that immortal light. A learned appreciates easy access of any educated person for being down to earth personality. A fool cannot judge the status and intellect thereby the philosophy of such an elevated soul. It is mistaken to recognize the learned as an ordinary man. It considers the right man in the wrong place. So it behaves wrongly.

Easy access has both value and no value. It gets either respect or no respect. He is ignored. Since he is easy he gets little respect. He may not ask for respect but he should be respected. Someone makes it easy to attract. A grocery shop falls in that category. A departmental store or an air-conditioned foreign bank does not offer easy access. Here only moneyed men get access. Poor persons avoid such elite establishments. Also such rich organizations restrict the entry of proletariat. Such organizations sell goods with the note of condition apply that remains quite illegible i.e., in a small font. This is just like hidden cost a tourist has to pay to the tour operator during or at the mid of the journey from where a tourist cannot return back home easily. Before journey behaviour was too easy and after commencement of journey it becomes too difficult. It is illegal. It harms good will the costly tool to attract customers. Only a reputed organization is cautious about it.

An easily got thing is either valued less or does not get proper recognition. Easy baby is neglected. Poor parents produce so many children easily. As such poor parents do not take proper care. Also they cannot nurture them properly. Many times they engage them in physical works. They are called child labourers. If the child is physically-handicapped then it is used in begging. The poor parents sometimes sell such handicapped children. The organized criminals purchase such children for business purpose. The mentally retarded children are killed since they are burden. If the mentally retarded child is a girl then it is kept for sexual abuse when grown up.

The poor parents have no money but can give birth to children so easily. The poor women are so fertile. But mortality rate is too high due to malnutrition of both mother and baby. Yet the poor woman produces baby untiringly till body permits. In contrast the rich parents are too infertile to give birth to a child. It seems money displaced the fertility. As such they adopt the children of poor parents. Sometimes they hire a surrogate woman. Thus those who have money have no children and those who have no money get children easily. Both the rich and the poor parents are cursed and in this respect both are equal and at par.

Bribe is called easy money. It is volatile in nature. One cannot use it properly. Rather wastage is its destiny. In contrast hard earned money is alias and akin to toil. Here bondage between person and penny is much. So a person honours every single penny. In case of dishonest income the bondage is very weak. So a dishonest person cannot resist the volatility of huge income. As such he becomes poor again. In contrast, hard earner can confirm his two times meal though hard earned money is limited. Through limited income he enjoys unlimited peace and happiness.

Only a lucky soul gets everything easily. A cursed soul nothing gets at ease. Even if he gets anything he cannot keep it. As such easy thing should be cared carefully.

Everything must be valued. A wise does it. He seldom cares whether he got it easily or through much effort. He preserves the thing properly. He minds it. He can use it as and when required.

Availability makes something easy. Abundance renders it cheap. In some places water is abundant. In some places it is scarce. But in both the cases it must be preserved so that at crisis period it is available. Generally, where it is available a person may not care it. At night when someone needs water and if he does not keep it beside the bed then he does not get it to quench his severe thirst. A wise knows it. He keeps the water bottle nearby. A fool does not preserve and die without water.

A person may be easy. But one should not consider all and everybody easy. If someone be easy may not be easy always. So better it is to care someone and show due and adequate respect. It confirms timely reciprocation. Generalization is not easy rather a risky game. It is the subject matter of a statistician.

Easy relation breaks easily. Bondage needs time. It should be time tested. Close relation may easily be far. Far relation may not be near at ease. It depends on the tackling power. An expert can convert a difficult relation into easy and easy relation into a difficult one through his innovative mechanism. All is not expert. So all cannot manage all and everybody as and when required. Generally, someone avoids a person when interest is served. Then the relation becomes nil due to out of sight out of mind.

Destruction is easy. One can destruct anything in the twinkle of an eye. Construction is difficult. It needs time. It demands perseverance. Rome was not built in a day. It is not easy to do anything overnight.

Easy achievement goes easy. Man does not mourn for this loss due to less or no attraction at all.

Honesty is the best policy. Honesty is singular in number like truth. It is the easiest avenue to realize objective. It is like to play in straight bat in cricket. Honesty causes no tension but pays immense return. None can raise any question while implementing it. Its methodology is straight forward i.e., as per rule. So deviation is not required.

Dishonesty means deviation from truth. It is alias and akin to lie.

Trust is an invisible currency, hard to earn, easy to lose.

In English grammar there are two articles viz., definite and indefinite. The is definite article. A and an are two indefinite articles. Definite article is single and singular in number. For example: The sun, the moon, etc. The sun and the moon are single. There are no other suns or no other moons in the solar system. Hence, to denote the singularity of the sun and the moon the definite article the is used before them.

In contrast there are so many stars in the sky. A star means any star of the sky. An apple means any apple of the tree. Both a and an are used to denote the indefinite status of anything. A and an highlight the plurality of something in question.

Truth is singular. It has single track. Lie may be single or plural in number. It may have one or many tracks. Sometimes it may be infinite in number. As such lying demands both intellect and memory. It is not an easy game. The liar must have to keep in mind the lie it uttered. If the liar tells different lies to different persons then it must have to keep in mind each lie and concerned persons accordingly. In contrast to speak the truth no home works are required. Here one must follow the rule. Here whims are not allowed. The track record of the truthful person is always too good and that of a liar is too bad. Track record stands for status.

He, who wants to deviate from the rule for self gain, has to explain the reason of deviation. Violation of rule causes punishment. As such they say public works are easier than private works. Public work follows single track. Private work follows different tracks. It changes the track in the mid-way for better gain. Gain is the single and sole agenda of any private organization. In every public work state gives protection. State is the guardian. In case of private venture each problem has to overcome by the concerned private organization.

Honesty needs no suppression. Dishonesty means to lie. To suppress fact or lie it needs much mental pressure. So lying is easy but suppression is difficult.

When something is easy it gives confidence. In the examination a mediocre student examines the first question and if it is easy then he writes the answer of that question. Then he considers the second question and if it is difficult then he omits it and proceeds towards the third question and so on.

In contrast a good student examines the whole question paper first. Then he separates both easy and difficult questions. He takes the risk of answering difficult questions there by impresses the examiner. Also the examiner gets relief from stereotype monotonous answers of the ordinary students. In fact a good student does not write different thing rather he writes differently thereby secures higher marks. This is the secret of good student and higher gradation.

Nothing is easy. Nothing is difficult. Nothing becomes easy automatically or spontaneously. Strong effort, uninterrupted devotion and sincerity render something easy. This is true in case of education, art, sports, acting, singing, dance, performance art, etc. i.e., in every sphere of life.

A student makes the difficult text easy studying continuously. A player becomes expert through regular practice. To an inattentive student lesson is not easy rather difficult. Similarly, a marginal player cannot score a single run. A sharp shooter can hit the target objet at ease. To attain this sharpness he has to practice day and night.

All is easy. All is difficult. What is easy to someone may be quite difficult to another one. Similarly, what is difficult to someone may be quite easy to another one. Thus, both easy and difficult are quite relative in nature. Here knack is the factor. A murderer cannot face an interview board. But he can murder easily. Similarly, the topper of the examination can face the interview board boldly. But he cannot murder at all. Rather he is afraid of knife and blood. His boldness disappears noticing bloodshed.

In the unknown environment everything is difficult. Only known face renders it easy. In the patriarchal society a newly-married bride finds in laws house difficult. Only love and affection of husband and in laws can make the life of bride easy. Love attracts. Trust binds. Cooperation confirms easy conjugal life.

Money makes life easy. Poverty renders it difficult. But money has its limitation. Money is a must but it is not all. The rich has the wrong notion that they can purchase anything and anybody. Anything is a commodity. Anybody is not or may not be a commodity. So question of easy purchase does not arise.

To a studious student lessons become easy. To an inattentive student syllabus is a burden. So easy is not an easy matter. It is not the spring that comes out from the hill spontaneously. One has to toil much to render it easy. Nothing, in this world, can be achieved overnight. They say Rome was not built in a day. Everything cannot be made up in stage. A wise knows it. A fool does not know it. As such it tries to make up everything in stage and fails easily and successfully.

Cheap may be easy. Easy may be cheap. Thus all cheap is easy but all easy may not be cheap. Easy is a blessing. Cheap is a curse. An elevated soul enjoys easy life. Cheap life means misery A wise knows it. A fool cannot distinguish between these two conditions of life. Cheap have three conditions. Firstly, it is poor in quality. Secondly, it does not last long. Lastly, it does not serve the purpose thereby seldom satisfies the need of the consumer or user.

CONCLUSION

It may be easy to marry a beautiful woman but difficult to hold her for long, just like easy earning money is not always easy to keep in safe custody. In both the cases it needs stamina. It demands intellect. Thus those who do lack in stamina and intellect as well and are poor both in appearance and wealth are well advised neither to dream for beautiful wife nor should they earn easy money. Yet man runs after beautiful woman and easy money whether they can hold it or not it matters little. It is greed that provokes man to achieve the desired thing. Man enjoys holding instinct and boasts of for the same. Holding is owner's pride, neighbours' envy. Failure ignites the craze more violently. Craze is an easy avenue to enjoy the life to the lees at ease and easily.

REFERENCES

No reference, since the present article is an outcome of Creative NonfictionWriting.

ABOUT THE AUTHOR

Pal, Dibakar is a Retired Executive Magistrate in India and PhD Student. Though he is a Civil Servant yet he is genuinely interested in diversified academic fields. As such, he holds master degrees in M.Sc(Math), M.A(English), M.A(Bengali), M.B.A(HRD), M.C.A, P.G.D.M.M(Marketing), L.L.B, D.C.E(Creative Writing), M. Phil (Business Management),UGC- NET(Management)-2008. He attended an International Conference at IIT, Mumbai, India and five International Conferences at U.S.A; though he gets invitation to present papers in many International Conferences at home and abroad round the year. He presents papers on Computer Science, Management, English Literature, Linguistic, Philosophy, Philology, Psychology, Sociology, Humanities and Poems. He presented a paper on Computer Science and Chaired in 2007 IEEE Conference at Richmond, Virginia, U.S.A. Also another paper on Fuzzy Logic was accepted by IEEE Conference 2010 at USA. He serves as Session Chair, Presider and Reviewer. He serves as reviewer of American Marketing Association, Journal of Common Ground; Australia, IEEE Transactions, IJEAPS, AJHC, Journal of Supercomputing.

He has written more than three hundred eighty five (385) articles in different fields. Among these one publication is as Monograph in International Journal on Management Science, one Monograph is in Journal of the World Universities Forum, one is in Consumer Behavior, two are in Computer Science, one is in Neuroscience, one is in Linguistic and rests are Creative Nonfiction Writing of English Literature. In Creative Nonfiction Writing two papers have been incorporated in SSRN's Top Ten Download List three times in November, December 2010 and April 2011. In Research Gate his papers have reached a milestone through more than 130000 reads. Scholars' Press and Lambert Academic Publishing House, Germany have published ten books on Creative Nonfiction, one book on Management Science, one book on Computer Science between the months July to December, 2016. New Texas, A Journal of Literature and Culture, Sul Ross State University, Alpine, Texas, USA has published ten Creative Nonfictions in February 2018. Notion Press, India has published a book on Creative Nonfiction in April 2020. Generis Publishing, USA has published one book on Creative Nonfiction in June 2020. International Educational Scientific Research Journal (E-ISSN : 2455-295X) publishes Creative Nonfiction every month regularly. Other publishers are Research Chronicler & Research Innovator (ISSN: 2347-503X; 2348-7674), SRYAHWA Publications, USA and IOJPH, Australia. Now he is pursuing his PhD thesis in Business Management in University of Calcutta, India. Also he is currently focussed on the Extension Works of Huffman Code i.e., Coding Theory and Pattern Recognition through Fuzzy Logic (Pattern Recognition, Image Processing, etc) of Computer Science.

His hobby is Creative Writing (Nonfiction). He says:

Creative writing is based more on manifestation rather than on expression. It does not inform, rather it reveals. So it bears no reference. The best creative writing is critical, and the best critical writing is creative. This article is an outcome of thinking about creative writing meant for a general readership. As such, I have adopted a free style methodology so that everyone can enjoy the pleasure of reading. As you might know, Francis Bacon (1561-1626), the immortal essayist, wrote many essays namely 'Of Love', 'Of Friendship', 'Of Ambition', 'Of Studies', and so on. The multiple-minded genius correctly pointed out that all the words of the dictionary can be used as themes for essays. But little has been done since his death to continue or finish his monumental task. Bacon's unique individual style of presentation ignited my imagination and encouraged me to write creative essays as a method of relieving a wide range of emotions through catharsis.

PUBLICATIONS OF BOOKS ON CREATIVE NONFICTION

[01]. OF HUMOUR AND SATIRE

[02]. CATHARSIS

[03]. SAID AND UNSAID

[04]. TOLD AND UNTOLD

[05]. UNTRODDEN HILL

[06]. FAR-OFF LAND

[07]. THEY SAY AND HEAR SAY

[08]. BOHEMIAN WHIMSICALITY

[09]. WESTERN HORIZON

[10]. HAPHAZARD

[11]. THOUSAND WEARY MILES

[12]. WOOD-LAND BLOSSOMS